Fundamental Biomechanics in Bone Tissue Engineering

Synthesis Lectures on Tissue Engineering

Editor
Kyriacos A. Athanasiou, *University of California, Davis*

The Synthesis Lectures on Tissue Engineering series will publish concise books on aspects of a field that holds so much promise for providing solutions to some of the most difficult problems of tissue repair, healing, and regeneration. The field of Tissue Engineering straddles biology, medicine, and engineering, and it is this multi-disciplinary nature that is bound to revolutionize treatment of a plethora of tissue and organ problems. Central to Tissue Engineering is the use of living cells with a variety of biochemical or biophysical stimuli to alter or maximize cellular functions and responses. However, in addition to its therapeutic potentials, this field is making significant strides in providing diagnostic tools.

Each book in the Series will be a self-contained treatise on one subject, authored by leading experts. Books will be approximately 65-125 pages. Topics will include 1) Tissue Engineering knowledge on particular tissues or organs (e.g., articular cartilage, liver), but also on 2) methodologies and protocols, as well as 3) the main actors in Tissue Engineering paradigms, such as cells, biomolecules, biomaterials, biomechanics, and engineering design. This Series is intended to be the first comprehensive series of books in this exciting area.

Fundamental Biomechanics in Bone Tissue Engineering

X. Wang, J.S. Nyman, X. Dong, H. Leng, and M. Reyes

ISBN: 978-3-031-01451-2   paperback
ISBN: 978-3-031-02579-2   ebook

DOI: 10.1007/978-3-031-02579-2

A Publication in the Springer series
SYNTHESIS LECTURES ON TISSUE ENGINEERING

Lecture #4
Series Editor: Kyriacos A. Athanasiou, *University of California, Davis*
Series ISSN
Synthesis Lectures on Tissue Engineering
Print 1944-0316   Electronic 1944-0308

Fundamental Biomechanics in Bone Tissue Engineering

X. Wang
University of Texas at San Antonio

J.S. Nyman
Vanderbilt University Medical Center

X. Dong
University of Texas at San Antonio

H. Leng
Peking University Third Hospital, China

M. Reyes
University of Texas at San Antonio

SYNTHESIS LECTURES ON TISSUE ENGINEERING #4

ABSTRACT

This eight-chapter monograph intends to present basic principles and applications of biomechanics in bone tissue engineering in order to assist tissue engineers in design and use of tissue-engineered products for repair and replacement of damaged/deformed bone tissues. Briefly, Chapter 1 gives an overall review of biomechanics in the field of bone tissue engineering. Chapter 2 provides detailed information regarding the composition and architecture of bone. Chapter 3 discusses the current methodologies for mechanical testing of bone properties (i.e., elastic, plastic, damage/fracture, viscoelastic/viscoplastic properties). Chapter 4 presents the current understanding of the mechanical behavior of bone and the associated underlying mechanisms. Chapter 5 discusses the structure and properties of scaffolds currently used for bone tissue engineering applications. Chapter 6 gives a brief discussion of current mechanical and structural tests of repair/tissue engineered bone tissues. Chapter 7 summarizes the properties of repair/tissue engineered bone tissues currently attained. Finally, Chapter 8 discusses the current issues regarding biomechanics in the area of bone tissue engineering.

KEYWORDS

tissue engineering, bone, biomechanics, mechanical properties, scaffold

Contents

Preface

Bone repair presents a unique challenge to tissue engineering strategies because bone defects often occur at sites that withstand significant mechanical loading. Thus, the design and fabrication of bone tissue engineering products often require both sufficient mechanical competence and adequate architecture that promotes osteogenesis. To help reconcile these opposing needs, this book provides basic knowledge on both the biomechanics of bone and the biomechanics of scaffolds currently employed in bone tissue engineering. The intent of this information is to assist tissue engineers not only in the design and fabrication of bone tissue engineering products, but also in the evaluation of such products and outcomes.

Tissue engineering is a multidisciplinary endeavor with progress occurring especially when there are advances in materials science, biomedical sciences, and engineering mechanics. This book takes the viewpoint of the latter discipline, specifically biomechanics, but offers findings and references from the other areas of research when appropriate. Thus, in addition to describing the biomechanical behavior of bone and scaffolds and their determinants, this book also discusses the cells and growth factors that stimulate osteogenesis as well as issues related to mechanobiology or the ability of bone cells to 'sense' changes in deformation (strain) and then to modify their behavior accordingly.

Our intent then is for the reader to understand bone as a biological material that adapts its structure and matrix quality to meet a mechanical function. In a sense, bone tissue engineering becomes a necessity when bone cannot do this for itself. A successful scaffold provides stability at the bone defect as new tissue forms and provides the stability. To achieve such a goal requires an understanding of bone composition, structure/architecture, and mechanical behavior. In providing this information, we pay particular attention to the difference between structure and material quality with emphasis on the hierarchical organization of bone because scaffolds have to 'bridge' the length scale of the bone cells to that of the defect. Moreover, for any given load, strain and failure depend on both structure/architecture and the inherent quality of native bone, scaffold, or regenerated bone tissue. Through choices in material synthesis and fabrication, there is some control over both of these attributes, and thus tissue engineering products can potentially be designed to target the mechanical stimuli for desired mechanotransduction events and to avoid failure of the product.

Educationally, this book can be used a reference book or a textbook for first year graduate students or senior undergraduate students in the program of biomedical engineering. There are numerous tables providing the biomechanical properties of various bones and repaired bone tissues. Also, testing methodologies are described in details for characterizing bone tissues and tissue engineering scaffolds.

Finally, we thank Dr. Rena Bizios for her constructive comments and other colleagues for providing figures for this book and valuable feed-back.

X. Wang, J.S. Nyman, X. Dong, H. Leng, and M. Reyes
December 2009

CHAPTER 1

Introduction

CHAPTER SUMMARY

This chapter provides an overview of the current status of bone tissue engineering and the role of biomechanics in this field. The objective and scope of this book are then discussed, followed by the organization of the monograph.

1.1 BACKGROUND

The function of human skeletal systems involves structurally supporting the body, protecting the vital organs, and serving as a reservoir of minerals for balanced metabolism. Clinically, most failures and/or defects of the skeletal system are induced by traumatic injuries, age-related or osteoporotic fractures, and pathological degenerations. Among the causes, age-related and osteoporotic fractures are increasingly becoming one of the major health care concerns around the world due to the increased risk of morbidity and the serious threat to the quality of life of patients. Recent studies have shown that the lifetime risk of major bone fractures is between 40-46% for Caucasian women and between 13-22% for Caucasian men at the age of 50 years [35, 44, 45, 56]. Furthermore, this number is rising every year and could reach 4.5 million world-wide by 2050 [28]. A similar trend can also be found in all other ethnic groups although fragility fracture rates are different [19,68]. In many such cases, surgical intervention with bone grafts and/or even total joint replacements are needed. Another common example of bone defects is congenital deformities, with functional and cosmetic corrections of these complications becoming a major clinical practice. The surgical procedures for such purposes primarily involve the transfer of tissues or the placement of implantable prosthesis. For example, the forehead flap of the skull has been used for local and subtotal nasal reconstruction. Moreover, distraction osteogenesis has been utilized in craniofacial surgeries for the correction of micrognathia [61], facial asymmetry [21,58], midface hypoplasia [70], and temporomandibular joint reconstruction [1,51].

The reconstructive surgery of failed and/or deformed skeletal systems is realized through replacing the defective tissues with viable, functioning ones. For minor fractures, bone is capable of self-regeneration within a few weeks without surgical interventions. For severe defects and/or deformities, bone grafting becomes necessary to restore its normal function because in this case bone cannot heal by itself. In the past, various techniques have been developed to stimulate the regeneration of new bone for better reconstruction of damaged tissues. A classical approach is the transplantation of homologous bone tissue to replace the damage one [16, 18, 22, 24, 33]. It involves the use of bone grafts (i.e., autografts or allografts) to provide the defect site with initial structural stability and osteogenic environment [29, 40, 86]. Autografts are bone segments taken from the patient's own body, whereas allografts are usually taken from cadaveric tissues. Although bone grafting has become a common practice in orthopaedic surgeries, both autografts and allografts are limited by certain uncontrollable factors. For autografts, the major shortcoming is the limited amount of bone stock that is available for harvesting [74]. Other considerations include the invasive nature of harvesting, in which the tissue is damaged and weakened at the harvest site, and the unpredictable resorption characteristics of the graft. The major limitation for allografts has been the immunogenic response to the foreign tissue of the graft, in which the implant is often rejected by the body and is subject to an inflammatory reaction [41, 50]. In addition, there is always a risk that allografts are capable of transmitting diseases and terminal sterilization techniques such as gamma irradiation can compromise the mechanical integrity of the allograft. Although a thorough screening process eliminates most of the grafts that have potential risks, it can not be guaranteed to exclude all contaminated ones [54]. These shortcomings have urged searching for more dependable sources for bone graft substitutes.

In past years, researchers from different disciplines (e.g., biomedical, biophysical, and biomaterials science/engineering, etc.) have collaborated closely to explore a new class of synthetic biomaterials that are capable of being implanted in the body to alleviate the aforementioned limitations induced by the conventional bone grafts. Based on the general definition, biomaterials are a class of materials that are systemically and pharmacologically inert, thus can be used for implantation within or incorporation with a living system. Nowadays, numerous synthetic bone graft materials have been developed to alleviate the practical complications associated with autografts and allografts [10, 48, 75, 83, 85]. Although good progress is being made, the function of these materials is quite different *in vivo* from that of natural bone tissues either compositionally or structurally. Current bone-replacement implants include ceramics, polymers, and some natural materials, such as collagen. Recently, engineering multi-phase materials (i.e., composites) with structure and composition similar to natural bone have been attempted. For example, nanocomposites, particularly hydroxyapatite (HA) and collagen-based, have gained much recognition as bone grafts due to their compositional and structural similarity with natural bone [39, 46, 49]. However, strict U.S. FDA guidelines have limited the use of such implants only to specific surgical applications. To date, no implant that mimics natural autogenous bone is clinically available. In fact, bone graft materials are not only required to be bioresorbable and to degrade into harmless byproducts that can be processed by the body, but also to stimulate new tissue generation.

Bone is a natural composite with a highly hierarchical structure, and it is a dynamic tissue capable of self-regenerating or self-remodeling throughout life. Thus, ideal synthetic bone grafts should have a broad range of properties and characteristics similar to those of natural bones and/or be engineered to stimulate or assist in tissue regeneration. This area of research is called tissue engineering and can be defined as the application of biological, chemical, and engineering principles toward the repair, restoration, or regeneration of living tissue by using scaffolds, cells, and growth factors either alone or in combination. By inducing the growth and regeneration of natural tissue, the implant function is no longer limited to a static role as a structural support, but it is capable of providing a suitable environment for bone ingrowth and can be completely replaced by natural tissues (biodegradable) over a desired time.

1.2   BONE TISSUE ENGINEERING

Tissue engineering is a fast growing scientific and technological field that aims at surgical repair or replacement of damaged/dysfunctional tissues by implanting cells, scaffolds, DNA, protein, and/or protein fragments. If successful, the limitations of existing bone grafting therapies can be completely circumvented. In bone tissue engineering, appropriate bioresorbable/biocompatible scaffold materials [5,38,72] may be combined with cells from various sources [43,55,84] and growth factors [37,76] to ensure adequate and timely tissue regeneration. Currently, there are three widely accepted approaches in tissue engineering: one is to use three-dimensional, porous, degradable scaffolds for providing mechanical support with repair dependent on the body itself for the recruitment and migration of host cells into the scaffold and differentiating into the desired tissue phenotype while replacing the degrading scaffold [12,67,79]. Another is to extract the appropriate cells from a patient, culture the cells *in vitro*, and then transplant the cultured cells back into the patient [52,60,64]. The last approach is to culture the cells of patients on a preformed three-dimensional scaffold *in vitro* and transplant the cell-scaffold construct into the patients [11,36,76]. Despite the early success of tissue engineering of soft tissues, tissue engineers still face challenges in repairing or replacing load-bearing tissues that serve a predominantly biomechanical function, such as bone and dental hard tissues. Progress in bone tissue engineering relies on a synergetic effort from a variety of areas of science, engineering, biomedicine, and biology. For example, the rapid achievements in this field have been the outcome of the recent significant advances in cell and molecular biology, e.g., the isolation and manipulation of cells, genes, and growth factors [53,59]. The advances in biomaterials have also provided new and innovative scaffold systems [17,63,78,81]. Furthermore, the integration of biology and materials science/engineering have allowed for delivering viable cells and for growing tissues on compatible constructs.

The major challenge in the research of tissue engineering is to create a favorable environment for the proliferation and differentiation of cells into functioning tissues. Three elements are inevitable for success of bone tissue engineering, namely the scaffold, the cells, and the environment in which the cell-scaffold constructs are cultured and conditioned.

First, optimization of the design and function of synthetic scaffolds is required to ensure the initial structural stability and to control the later tissue regeneration process. Numerous scaffold properties may be manipulated, including the type of material utilized, architecture of scaffolds (e.g., the shape and size of pores in which cells are placed), the mechanical properties of the construct, surface coating aimed towards promotion of cell adhesion, and incorporation of chemicals and growth factors conducive to tissue formation.

Next, adequate cells with osteogenic potential are also a key element for successful tissue regeneration following the selection of scaffolds with desirable characteristics. The cells may be obtained from the patient (autologous), another human donor (allogeneic), and another species (xenogeneic). Autologous cells are the most desirable in terms of legal and immune rejection issues. However, these cells may be unhealthy and AGEs accumulation in bone may affect both the structural integrity of bone and the biological process during bone resorption. All other cell sources certainly present problems with regards to immune rejection and other genetic issues. Only if genetic engineering approaches can resolve this tissue, can they be used for human tissue engineering. Currently, four cell types have been considered for bone tissue engineering applications: bone marrow cells [14,73], mesenchymal stem cells [25,52], muscle cells [7,26], and embryonic stem cells [30,77].

Finally, conditioning of cell-seeded constructs also plays an important role in control and optimization of tissue production. Currently, it is achieved *in vitro* using so-called bioreactors. At present, development of bioreactors is still in experimental stages and the practical use of them is in progress. Generally speaking, multiple functions are required for bioreactors to supply nutrients and dissolvable gases, to remove the waste from cells and degradation products of the scaffold, to maintain an appropriate physical environment (e.g., fluid flow, mechanical stress/strain, etc.), and to provide needed growth factors for controlled cell proliferation and differentiation. All of those will occur within a porous three-dimensional scaffold controlled by the bioreactor. Thus, the selection of a bioreactor is dependent on the specific tissue type and the selected scaffold.

1.3   BIOMECHANICS VS. BONE TISSUE ENGINEERING

Skeletal tissues mainly serve biomechanical functions. Thus, the biomechanical properties of the tissues are critical to their proper performance *in vivo*. In order to repair or replace damaged load-bearing tissues both efficiently and effectively, the following issues have to be addressed [9].

First, it is important to know the physiological ranges of forces, stresses, and strains that normal bone tissues transmit or encounter during daily activities [3, 71, 80]. This information is critical for tissue engineers to design the tissue engineered constructs that could adapt to the mechanical environment for optimized outcome of the repair. From prior experimental measurements involving humans, *in vivo* bone strains under different physiological conditions have been reported [8,62,65].

Next, it is necessary to understand the *in vivo* mechanical behavior of bone tissues when subjected to expected stresses and strains, as well as under failure conditions. Based on this understanding, tissue engineers could identify the most important properties that should be incorporated into the optimal design of synthetic scaffolds and/or cell-scaffold constructs in order to sustain the structural stability of the implants. To date, numerous data in this regard are available in the literature for different tissue types and anatomic locations.

After the repair process, the mechanical properties of the regenerated tissues should be evaluated to ensure whether or not the tissue engineered bone tissues are good enough to sustain the functions of normal tissues. However, it is still a debatable issue whether tissue engineered bone and replacement implants need to exactly duplicate the structure and function of the normal bone tissues.

Moreover, the role of physical factors, such as local mechanical stress that regulate cell behavior, should also be considered in the design of tissue engineered constructs (implants). It has been reported that bone cells respond even to small deformation applied to the bulk tissue [6, 13, 31, 42]. Thus, control of deformation in the scaffold is critical for bone cells to sense changes in the surrounding mechanical environment thereby modulating their activity accordingly. Ultimately, these cells are responsible for maintaining mechanostasis of the cell-scaffold constructs depending on the strain placed upon the structure. In addition, bone formation and remodeling around implanted constructs is influenced by the loading environment. Thus, the surface properties of implant material and the anatomical site of implantation are also key elements to successful implant integration.

Finally, tissue engineers need to understand the mechanobiology of bone since the mechanical stimuli to these implants both *in vitro* and *in vivo* can affect the outcome of repair. Convincing evidence of the importance of mechanical stimuli to bone regeneration is distraction osteogenesis, whose principle has been widely applied to bone lengthening therapies in clinical practice [2,23,34]. In addition, it has been practiced in hospitals for a long time to subject broken limbs to traction for better healing. Without such treatments, bone repair may become incorrect, causing misshapen bones or limbs. In fact, cellular components of bone play a significant role in mechanotransduction for activation and control of metabolism during bone resorption/formation and remodeling processes. Mechanical loading at physiologically-relevant magnitudes has been shown to impose a significant effect on bone deposition and remodeling in human bone [4,47,57,66], in animal models [15,27,32], and in cell cultures [69]. On the contrary, bone loss is induced if no appropriate loading is applied [82]. In fact, bone remodeling is accomplished by synchronized and/or sequential actions of multiple cells (i.e., osteoprogenitor cells, osteoblasts, osteocytes, osteoclasts, and bone-lining cells). The cell system is reacting to external mechanical stimuli by building the bone matrix, maintenance of the tissue, and remodeling as required, respectively. Thus, understanding mechanotransduction pathways of bone cells is critical to successful tissue engineering of bone grafts. In addition, it is important to know how physical factors influence cellular activity in bioreactors and whether or not cell-scaffold constructs that are mechanically stimulated before surgery are desired to produce a better outcome.

Combining the aforementioned knowledge, tissue engineers are required to objectively incorporate these functional criteria in the design, manufacture, and optimization of tissue engineered constructs. For example, three approaches for translating cellular biomechanics into the field of bone tissue engineering have been proposed by El Haj et al. [20]: (1) design of coatings on the scaffold surfaces for optimized cell adhesion and signal translation of load; (2) development of 'mechano-active' scaffolds, where controlled release of a chemical agonist is incorporated to attenuate activation times of voltage-operated calcium channels, thus regulating the mechanical response of cells; and (3) optimizing biomechanical environment in bioreactors based on the desired properties and architecture of the tissue engineered bone. Obviously, the strategies for manufacturing tissue-engineered products will be unique to each individual case under consideration. Thus, what constitutes "success" will vary among individual applications. For example, implants that are designed to replace load-bearing tissues may require a higher strength and toughness than those that are designed to improve the quality of life, such as cosmetic corrections.

1.4   OBJECTIVES AND SCOPES

In this monograph, the authors intend to present basic principles and applications of biomechanics in bone tissue engineering in order to help tissue engineers in the design and use of tissue-engineered constructs that repair and replace of both non load-bearing and load-bearing bone tissues. First, to understand bone mechanics, one needs to know the composition and architecture of bone tissues because such information is the basis for understanding the mechanical behavior of the tissue. In addition, tissue engineers need to know *in vivo* stress/strain histories for a variety of physical activities of humans since these *in vivo* data provide mechanical thresholds that tissue repairs/replacements will likely encounter after surgery. Moreover, such information is extremely important in connecting bone mechanical and structural properties to the cellular activities during the formation process of bone tissues. Secondly, the structure-function relationship of bone tissues needs to be established in both macroscopic and micro/nanoscopic levels. Obviously, the required mechanical properties must be established under normal physiological and failure conditions for the tissue-engineered products. Based on these baseline data, tissue engineers may determine design parameters within the expected thresholds for different *in vivo* activities and safety factors for unexpected situations beyond the threshold levels. Thirdly, given that the mechanical properties of the designs are not expected to completely duplicate the properties of the native tissues, a subset of the mechanical properties must be selected and prioritized. This will ensure cost-effective approaches for design and manufacturing of tissue engineered bone products. Fourthly, test standards must be set when evaluating the repairs/replacements after surgery so as to determine whether these treatments are successful or not. For example, desired mechanical characteristics of the repairs and replacements cannot necessarily guarantee that other aspects of the repair outcome are also satisfactory. Thus, new and improved methods must also be developed for assessing the function of engineered tissues. Finally, the effects of physical factors on cellular activity must be determined in engineered tissues. Knowing the underlying mechanisms of the effects may facilitate design of multi-functional scaffolds and bioreactors to direct cellular activity and phenotype toward a desired end goal. Combining these principles into functional tissue engineering should result in development of strategies for safer and more efficacious repairs and replacements of bone tissues.

1.5   ORGANIZATION OF THE MONOGRAPH

In this monograph, Chapter 1 introduces the field of bone tissue engineering and its relationship with biomechanics. Chapter 2 discusses bone composition and architecture in detail in order to provide a structural basis for understanding bone mechanics in bone tissue engineering. Chapter 3 discusses the current methodologies for mechanical testing of bone tissues in all aspects: elastic, plastic, damage/fracture mechanics, and viscoelastic/viscoplastic properties. Chapter 4 presents the mechanical behavior of bone tissues and the current understanding of underlying mechanisms related to such behaviors. Chapter 5 discusses the structure and properties of scaffolds currently used for bone tissue engineering applications. Chapter 6 gives a brief discussion of current mechanical and structural tests of repair/tissue engineered bone tissues. Chapter 7 provides the properties of repair/tissue engineered bone tissues obtained using current methodologies. Finally, Chapter 8 presents the current biomechanics issues in bone tissue engineering.

REFERENCES

[1] Abbas, I., et al., Temporomandibular joint ankylosis: experience with interpositional gap arthroplasty at Ayub Medical College Abbottabad. Journal of Ayub Medical College, Abbottabad: JAMC, 2005. **17**(4): p. 67–9. 1.1

[2] Aronson, J., Experimental and clinical experience with distraction osteogenesis. The Cleft palate-craniofacial journal, 1994. **31**(6): p. 473–81; DOI: 10.1597/1545-1569(1994)031%3C0473:EACEWD%3E2.3.CO;2 1.3

[3] Asundi, A. and A. Kishen, A strain gauge and photoelastic analysis of in vivo strain and in vitro stress distribution in human dental supporting structures. Archives of oral biology, 2000. **45**(7): p. 543–50. DOI: 10.1016/S0003-9969(00)00031-5 1.3

[4] Bakker, A., J. Klein-Nulend, and E. Burger, Shear stress inhibits while disuse promotes osteocyte apoptosis. Biochemical and biophysical research communications, 2004. **320**(4): p. 1163–8. DOI: 10.1016/j.bbrc.2004.06.056 1.3

[5] Boccaccini, A.R. and J.J. Blaker, Bioactive composite materials for tissue engineering scaffolds. Expert review of medical devices, 2005. **2**(3): p. 303–17. DOI: 10.1586/17434440.2.3.303 1.2

[6] Bonewald, L.F., Osteocytes: a proposed multifunctional bone cell. Journal of musculoskeletal & neuronal interactions, 2002. **2**(3): p. 239–41. 1.3

[7] Bueno, D.F., et al., New Source of Muscle-Derived Stem Cells with Potential for Alveolar Bone Reconstruction in Cleft Lip and/or Palate Patients. Tissue engineering. Part A, 2009. **15**(2): p. 427–35. DOI: 10.1089/ten.tea.2007.0417 1.2

[8] Burr, D.B., et al., In vivo measurement of human tibial strains during vigorous activity. Bone, 1996. **18**(5): p. 405–10. DOI: 10.1016/8756-3282(96)00028-2 1.3

[9] Butler, D.L., S.A. Goldstein, and F. Guilak, Functional Tissue Engineering: The Role of Biomechanics. Journal of biomechanical engineering, 2000. **122**(6): p. 570–575. DOI: 10.1115/1.1318906 1.3

[10] Carmagnola, D., et al., Oral implants placed in bone defects treated with Bio-Oss, Ostim-Paste or PerioGlas: an experimental study in the rabbit tibiae. Clinical oral implants research, 2008. **19**(12): p. 1246–53. DOI: 10.1111/j.1600-0501.2008.01584.x 1.1

[11] Carstens, M.H., M. Chin, and X.J. Li, In situ osteogenesis: regeneration of 10-cm mandibular defect in porcine model using recombinant human bone morphogenetic protein-2 (rhBMP-2) and Helistat absorbable collagen sponge. The Journal of craniofacial surgery, 2005. **16**(6): p. 1033–42. DOI: 10.1097/01.scs.0000186307.09171.20 1.2

[12] Chen, F., et al., Segmental bone tissue engineering by seeding osteoblast precursor cells into titanium mesh-coral composite scaffolds. International journal of oral and maxillofacial surgery, 2007. **36**(9): p. 822–7. DOI: 10.1016/j.ijom.2007.06.019 1.2

[13] Cherian, P.P., et al., Mechanical strain opens connexin 43 hemichannels in osteocytes: a novel mechanism for the release of prostaglandin. Molecular biology of the cell, 2005. **16**(7): p. 3100–6. DOI: 10.1091/mbc.E04-10-0912 1.3

[14] Connolly, J.F., Injectable bone marrow preparations to stimulate osteogenic repair. Clinical orthopaedics and related research, 1995(313): p. 8–18. 1.2

[15] Cowin, S.C., et al., Functional adaptation in long bones: establishing in vivo values for surface remodeling rate coefficients. Journal of biomechanics, 1985. **18**(9): p. 665–84. DOI: 10.1016/0021-9290(85)90022-3 1.3

[16] Cutter, C.S. and B.J. Mehrara, Bone grafts and substitutes. Journal of long-term effects of medical implants, 2006. **16**(3): p. 249–60. 1.1

[17] Dickson, G., et al., Orthopaedic tissue engineering and bone regeneration. Technology and health care: official journal of the European Society for Engineering and Medicine, 2007. **15**(1): p. 57–67. 1.2

[18] Dragan, S., et al., Our experiences in the reconstruction of bone stock deficiency with allogenous lyophilized bone graft after surgical resection of the giant-cell tumour. Ortopedia, traumatologia, rehabilitacja, 2003. **5**(4): p. 508–17. 1.1

[19] Duan, Y., et al., Structural and biomechanical basis of racial and sex differences in vertebral fragility in Chinese and Caucasians. Bone., 2005. **36**(6): p. 987–98. DOI: 10.1016/j.bone.2004.11.016 1.1

[20] El Haj, A.J., et al., Controlling cell biomechanics in orthopaedic tissue engineering and repair. Pathologie-biologie, 2005. **53**(10): p. 581–9. DOI: 10.1016/j.patbio.2004.12.002 1.3

[21] Ellis, E., 3rd and D.P. Sinn, Use of homologous bone in maxillofacial surgery. Journal of oral and maxillofacial surgery: official journal of the American Association of Oral and Maxillofacial Surgeons, 1993. **51**(11): p. 1181–93. 1.1

[22] Enneking, W.F., J.L. Eady, and H. Burchardt, Autogenous cortical bone grafts in the reconstruction of segmental skeletal defects. The Journal of bone and joint surgery. American volume, 1980. **62**(7): p. 1039–58. 1.1

[23] Erdem, M., et al., Lengthening of short bones by distraction osteogenesis–results and complications. International orthopaedics, 2009. **33**(3): p. 807–13. DOI: 10.1007/s00264-007-0491-x 1.3

[24] Ferretti, C. and U. Ripamonti, Human segmental mandibular defects treated with naturally derived bone morphogenetic proteins. The Journal of craniofacial surgery, 2002. **13**(3): p. 434–44. DOI: 10.1097/00001665-200205000-00014 1.1

[25] Filho Cerruti, H., et al., Allogenous bone grafts improved by bone marrow stem cells and platelet growth factors: clinical case reports. Artificial organs, 2007. **31**(4): p. 268–73. DOI: 10.1111/j.1525-1594.2007.00374.x 1.2

[26] Goodell, M.A., et al., Stem cell plasticity in muscle and bone marrow. Annals of the New York Academy of Sciences, 2001. **938**: p. 208–18; discussion 218–20. DOI: 10.1111/j.1749-6632.2001.tb03591.x 1.2

[27] Gross, T.S., et al., Strain gradients correlate with sites of periosteal bone formation. Journal of bone and mineral research: the official journal of the American Society for Bone and Mineral Research, 1997. **12**(6): p. 982–8. 1.3

[28] Gullberg, B., O. Johnell, and J.A. Kanis, World-wide projections for hip fracture. Osteoporosis international: a journal established as result of cooperation between the European Foundation for Osteoporosis and the National Osteoporosis Foundation of the USA, 1997. **7**(5): p. 407–13. 1.1

[29] Haidukewych, G.J. and D.J. Berry, Nonunion of fractures of the subtrochanteric region of the femur. Clinical orthopaedics and related research, 2004(419): p. 185–8. 1.1

[30] Handschel, J., et al., Induction of osteogenic markers in differentially treated cultures of embryonic stem cells. Head & face medicine, 2008. **4**: p. 10. 1.2

[31] Harter, L.V., K.A. Hruska, and R.L. Duncan, Human osteoblast-like cells respond to mechanical strain with increased bone matrix protein production independent of hormonal regulation. Endocrinology, 1995. **136**(2): p. 528–35. DOI: 10.1210/en.136.2.528 1.3

[32] Hazenberg, J.G., et al., Microdamage: a cell transducing mechanism based on ruptured osteocyte processes. Journal of biomechanics, 2006. **39**(11): p. 2096–103. DOI: 10.1016/j.jbiomech.2005.06.006 1.3

[33] Hoffman, H.T., et al., Mandible reconstruction with vascularized bone grafts. A histologic evaluation. Archives of otolaryngology–head & neck surgery, 1991. **117**(8): p. 917–25. 1.1

[34] Hönig, J.F., U.A. Grohmann, and H.A. Merten, Mandibula distraction osteogenesis for lengthening the mandibula to correct a malocclusion: a more than 70-year-old German

concept in craniofacial surgery. The Journal of craniofacial surgery, 2002. **13**(1): p. 96–8. DOI: 10.1097/00001665-200201000-00022 1.3

[35] Hordon, L.D., et al., Trabecular architecture in women and men of similar bone mass with and without vertebral fracture: I. Two-dimensional histology. Bone, 2000. **27**(2): p. 271–6. DOI: 10.1016/S8756-3282(00)00329-X 1.1

[36] Howard, D., et al., Tissue engineering: strategies, stem cells and scaffolds. Journal of anatomy, 2008. **213**(1): p. 66–72. DOI: 10.1111/j.1469-7580.2008.00878.x 1.2

[37] Huang, Y.C., et al., Combined angiogenic and osteogenic factor delivery enhances bone marrow stromal cell-driven bone regeneration. Journal of bone and mineral research: the official journal of the American Society for Bone and Mineral Research, 2005. **20**(5): p. 848–57. 1.2

[38] Hutmacher, D.W., Scaffolds in tissue engineering bone and cartilage. Biomaterials, 2000. **21**(24): p. 2529–43. DOI: 10.1016/S0142-9612(00)00121-6 1.2

[39] Itoh, S., et al., Development of a hydroxyapatite/collagen nanocomposite as a medical device. Cell transplantation, 2004. **13**(4): p. 451–61. DOI: 10.3727/000000004783983774 1.1

[40] Jaffe, K.A., E.P. Launer, and B.M. Scholl, Use of a fibular allograft strut in the treatment of benign lesions of the proximal femur. American journal of orthopedics (Belle Mead, N.J.), 2002. **31**(10): p. 575–8. 1.1

[41] Janssen, M.E., C. Lam, and R. Beckham, Outcomes of allogenic cages in anterior and posterior lumbar interbody fusion. European spine journal: official publication of the European Spine Society, the European Spinal Deformity Society, and the European Section of the Cervical Spine Research Society, 2001. **10**: p. S158–68. DOI: 10.1007/s005860100292 1.1

[42] Jones, D.B., et al., Biochemical signal transduction of mechanical strain in osteoblast-like cells. Biomaterials, 1991. **12**(2): p. 101–10. DOI: 10.1016/0142-9612(91)90186-E 1.3

[43] Kanczler, J.M. and R.O. Oreffo, Osteogenesis and angiogenesis: the potential for engineering bone. European cells & materials, 2008. **15**: p. 100–14. 1.2

[44] Kanis, J.A., et al., Risk of hip fracture derived from relative risks: an analysis applied to the population of Sweden. Osteoporosis international: a journal established as result of cooperation between the European Foundation for Osteoporosis and the National Osteoporosis Foundation of the USA, 2000. **11**(2): p. 120–7. 1.1

[45] Kanis, J.A., et al., Risk of hip fracture according to the World Health Organization criteria for osteopenia and osteoporosis. Bone, 2000. **27**(5): p. 585–90. DOI: 10.1016/S8756-3282(00)00381-1 1.1

[46] Kaufman, J.D., J. Song, and C.M. Klapperich, Nanomechanical analysis of bone tissue engineering scaffolds. Journal of biomedical materials research. Part A, 2007. **81**(3): p. 611–23. DOI: 10.1002/jbm.a.30976 1.1

[47] Keller, T.S. and A.M. Strauss, Predicting skeletal adaptation in altered gravity environments. Journal of the British Interplanetary Society, 1993. **46**(3): p. 87–96. 1.3

[48] Kim, H.W., et al., Bone formation on the apatite-coated zirconia porous scaffolds within a rabbit calvarial defect. Journal of biomaterials applications, 2008. **22**(6): p. 485–504. DOI: 10.1177/0885328207078075 1.1

[49] Kim, K. and J.P. Fisher, Nanoparticle technology in bone tissue engineering. Journal of drug targeting, 2007. **15**(4): p. 241–52. DOI: 10.1080/10611860701289818 1.1

[50] Kocher, M.S., M.C. Gebhardt, and H.J. Mankin, Reconstruction of the distal aspect of the radius with use of an osteoarticular allograft after excision of a skeletal tumor. The Journal of bone and joint surgery. American volume, 1998. **80**(3): p. 407–19. 1.1

[51] Kudo, K., et al., Surgical correction and osteoplasty for forward dislocation of temporomandibular joint. The Journal of Nihon University School of Dentistry, 1989. **31**(4): p. 577–84. 1.1

[52] Kulakov, A.A., et al., Clinical study of the efficiency of combined cell transplant on the basis of multipotent mesenchymal stromal adipose tissue cells in patients with pronounced deficit of the maxillary and mandibulary bone tissue. Bulletin of experimental biology and medicine, 2008. **146**(4): p. 522–5. DOI: 10.1007/s10517-009-0322-8 1.2

[53] Leonardi, E., et al., Isolation, characterisation and osteogenic potential of human bone marrow stromal cells derived from the medullary cavity of the femur. La Chirurgia degli organi di movimento, 2008. **92**(2): p. 97–103. DOI: 10.1007/s12306-008-0057-0 1.2

[54] Lo Monte, A.I., et al., The new immunosuppressive era. Annali italiani di chirurgia, 1999. **70**(1): p. 1–19. 1.1

[55] Malicev, E., et al., Growth and differentiation of alveolar bone cells in tissue-engineered constructs and monolayer cultures. Biotechnology and bioengineering, 2008. **100**(4): p. 773–81. DOI: 10.1002/bit.21815 1.2

[56] Melton, L.J., 3rd, et al., Risk of age-related fractures in patients with primary hyperparathyroidism. Archives of internal medicine, 1992. **152**(11): p. 2269–73. 1.1

[57] Mittlmeier, T., C. Mattheck, and F. Dietrich, Effects of mechanical loading on the profile of human femoral diaphyseal geometry. Medical engineering & physics, 1994. **16**(1): p. 75–81. DOI: 10.1016/1350-4533(94)90014-0 1.3

[58] Munro, I.R., Rigid fixation and facial asymmetry. Clinics in plastic surgery, 1989. **16**(1): p. 187–94. 1.1

[59] Park, D.H., et al., From the basics to application of cell therapy, a steppingstone to the conquest of neurodegeneration: a meeting report. Medical science monitor: international medical journal of experimental and clinical research, 2009. **15**(2): p. RA23–31. 1.2

[60] Peng, H. and J. Huard, Muscle-derived stem cells for musculoskeletal tissue regeneration and repair. Transplant immunology, 2004. **12**(3–4): p. 311–9. 1.2

[61] Pensler, J.M., R.D. Christopher, and D.C. Bewyer, Correction of micrognathia with ankylosis of the temporomandibular joint in childhood. Plastic and reconstructive surgery, 1993. **91**(5): p. 799–805. 1.1

[62] Perusek, G.P., et al., An extensometer for global measurement of bone strain suitable for use in vivo in humans. Journal of biomechanics, 2001. **34**(3): p. 385–91. DOI: 10.1016/S0021-9290(00)00197-4 1.3

[63] Pound, J.C., et al., An ex vivo model for chondrogenesis and osteogenesis. Biomaterials, 2007. **28**(18): p. 2839–49. DOI: 10.1016/j.biomaterials.2007.02.029 1.2

[64] Redlich, A., et al., Bone engineering on the basis of periosteal cells cultured in polymer fleeces. Journal of materials science. Materials in medicine, 1999. **10**(12): p. 767–72. DOI: 10.1023/A:1008994715605 1.2

[65] Rolf, C., et al., An experimental in vivo method for analysis of local deformation on tibia, with simultaneous measures of ground reaction forces, lower extremity muscle activity and joint motion. Scandinavian journal of medicine & science in sports, 1997. **7**(3): p. 144–51. DOI: 10.1111/j.1600-0838.1997.tb00131.x 1.3

[66] Ruff, C.B., W.W. Scott, and A.Y. Liu, Articular and diaphyseal remodeling of the proximal femur with changes in body mass in adults. American journal of physical anthropology, 1991. **86**(3): p. 397–413. DOI: 10.1002/ajpa.1330860306 1.3

[67] Schantz, J.T., et al., Repair of calvarial defects with customized tissue-engineered bone grafts I. Evaluation of osteogenesis in a three-dimensional culture system. Tissue engineering, 2003. **9**: p. S113–26. DOI: 10.1089/10763270360697021 1.2

[68] Schnitzler, C.M., Bone quality: a determinant for certain risk factors for bone fragility. Calcif Tissue Int, 1993. **53**(Suppl 1): p. S27–31. DOI: 10.1007/BF01673398 1.1

[69] Scott, A., et al., Mechanotransduction in human bone: in vitro cellular physiology that underpins bone changes with exercise. Sports medicine (Auckland, N.Z.), 2008. **38**(2): p. 139–60. DOI: 10.2165/00007256-200838020-00004 1.3

[70] Seyhan, T., B.H. Kircelli, and B. Caglar, Correction of septal and midface hypoplasia in maxillonasal dysplasia (Binder's syndrome) using high-density porous polyethylene. Aesthetic plastic surgery, 2009. **33**(4): p. 661–5. DOI: 10.1007/s00266-009-9312-5 1.1

[71] Sharkey, N.A., et al., Strain and loading of the second metatarsal during heel-lift. The Journal of bone and joint surgery. American volume, 1995. **77**(7): p. 1050–7. 1.3

[72] Shikinami, Y., et al., Bioactive and bioresorbable cellular cubic-composite scaffolds for use in bone reconstruction. Journal of the Royal Society, Interface / the Royal Society, 2006. **3**(11): p. 805–21. DOI: 10.1098/rsif.2006.0144 1.2

[73] Soltan, M., D. Smiler, and J.H. Choi, Bone marrow: orchestrated cells, cytokines, and growth factors for bone regeneration. Implant dentistry, 2009. **18**(2): p. 132–41. DOI: 10.1097/ID.0b013e3181990e75 1.2

[74] Springfield, D., Autograft reconstructions. The Orthopedic clinics of North America, 1996. **27**(3): p. 483–92. 1.1

[75] Stubbs, D., et al., In vivo evaluation of resorbable bone graft substitutes in a rabbit tibial defect model. Biomaterials, 2004. **25**(20): p. 5037–44. DOI: 10.1016/j.biomaterials.2004.02.014 1.1

[76] Tabata, Y., Tissue regeneration based on growth factor release. Tissue engineering, 2003. **9**: p. S5–15. DOI: 10.1089/10763270360696941 1.2

[77] Tian, X.F., et al., Comparison of osteogenesis of human embryonic stem cells within 2D and 3D culture systems. Scandinavian journal of clinical and laboratory investigation, 2008. **68**(1): p. 58–67. DOI: 10.1080/00365510701466416 1.2

[78] Trojani, C., et al., Three-dimensional culture and differentiation of human osteogenic cells in an injectable hydroxypropylmethylcellulose hydrogel. Biomaterials, 2005. **26**(27): p. 5509–17. DOI: 10.1016/j.biomaterials.2005.02.001 1.2

[79] Tyler, M.S. and D.P. McCobb, The genesis of membrane bone in the embryonic chick maxilla: epithelial-mesenchymal tissue recombination studies. Journal of embryology and experimental morphology, 1980. **56**: p. 269–81. 1.2

[80] Van Rietbergen, B., et al., Tissue stresses and strain in trabeculae of a canine proximal femur can be quantified from computer reconstructions. Journal of biomechanics, 1999. **32**(4): p. 443–51. 1.3

[81] Vaquette, C., et al., An innovative method to obtain porous PLLA scaffolds with highly spherical and interconnected pores. Journal of biomedical materials research. Part B, Applied biomaterials, 2008. **86**(1): p. 9–17. DOI: 10.1002/jbm.b.30982 1.2

[82] Verhaeghe, J., et al., Effects of exercise and disuse on bone remodeling, bone mass, and biomechanical competence in spontaneously diabetic female rats. Bone, 2000. **27**(2): p. 249–56. DOI: 10.1016/S8756-3282(00)00308-2 1.3

[83] Walsh, W.R., et al., Beta-TCP bone graft substitutes in a bilateral rabbit tibial defect model. Biomaterials, 2008. **29**(3): p. 266–71. DOI: 10.1016/j.biomaterials.2007.09.035 1.1

[84] Xiao, Y., et al., Tissue engineering for bone regeneration using differentiated alveolar bone cells in collagen scaffolds. Tissue engineering, 2003. **9**(6): p. 1167–77. DOI: 10.1089/10763270360728071 1.2

[85] Yu, H., et al., Effect of porosity and pore size on microstructures and mechanical properties of poly-epsilon-caprolactone- hydroxyapatite composites. Journal of biomedical materials research. Part B, Applied biomaterials, 2008. **86B**(2): p. 541–7. DOI: 10.1002/jbm.b.31054 1.1

[86] Zatsepin, S.T. and V.N. Burdygin, Replacement of the distal femur and proximal tibia with frozen allografts. Clinical orthopaedics and related research, 1994(303): p. 95–102. 1.1

CHAPTER 2

Bone Composition and Structure

CHAPTER SUMMARY

The major tasks of bone tissue engineering are to fabricate functional scaffold systems that can mimic the natural tissue and to ensure the successful regeneration of defect tissue to recover its normal functions. To achieve the goals, it is necessary to understand the composition and structure of bone. This chapter presents a detailed description of bone as a natural and highly hierarchical composite material in order to provide the information that tissue engineers require in the design and fabrication of functional scaffold systems and in the evaluation of tissue engineered/repair bone tissues. In addition, differences in the composition and structure of bone between species are also discussed in the event that the information is needed by tissue engineers and researchers for choosing adequate animal models in both basic and clinical research on bone tissue engineering.

2.1 INTRODUCTION

The major objective of bone tissue engineering is to provide proper bone grafts or constructs for successful replacement/repair of defect bone tissues [1]. Its application is two-fold: (1) synthetic bone grafts for replacement of defect tissue and (2) cell-scaffold constructs for repair of bone fractures. In order to ensure tissue engineered bone grafts/implants function well, they should mimic natural bone tissues and satisfy the mechanical and biological requirements in replacement/repair of the defect tissues [2] [3]. For example, mechanical stability is a major concern pertaining to load bearing capacity of repaired tissue using orthopaedic implant systems. To deal with the challenge, it is necessary to know how the mechanical performance of bone is related to its architecture and composition. In addition, the biological environment directly regulates the process of bone formation and subsequently affects the architecture/composition of tissue. Thus, understanding the structure and composition of bone is one of the prerequisites for meaningful tissue-engineering research and clinical practices.

   Anatomically, tissue engineers have to take into consideration two important aspects in research and design of bone tissue engineering products. One is the architecture of bone that plays a major role in the structural integrity of bones (e.g., stiffness or rigidity), while the other is the composition/ultrastructure of bone that mainly determines the material behavior of the tissue (e.g., post-yielding and failure, etc.). For example, the porosity, trabecular connectivity, and trabecular thickness of cancellous bone are the major factors that affect the structural stability of this type of tissue. On the other hand, the mineral density, collagen integrity, and the interactions of the two phases of bone govern the tissue behavior during the deformation of the tissue. Also important is that both architecture and ultrastructure/composition of bone are dependent on multiple factors, such as age, gender, and anatomic locations.

Another important issue for bioengineers to keep in mind is the difference in bone architecture and composition among species. The issue arises from the fact that animal models have been most often used in tissue engineering research for preclinical trials in order to develop new bone tissue engineering techniques and products. All newly developed bone grafts are tested in animals before clinical trials in humans to verify their efficacy in serving the prescribed functions [4]. Compared with humans, animals are much easier to handle and control in research. Although bone tissues from some animal species (e.g., primates, canine, etc.) are similar in composition and structure [5–8], most animal models currently used in bone tissue engineering research (i.e., mice and rats) are substantially different from humans. For example, human cortical bone is osteonal in nature, whereas murine cortical bone lacks osteons with the lamellae running circumferentially around the medullary canal. Therefore, understanding their differences is important to interpreting results from an animal study involving bone-tissue-engineering research [9].

In this chapter, we provide information on the compositional and architectural properties of bone and their contribution to the mechanical behavior of bone, which is important to the appropriate design and manufacturing of synthetic bone grafts for human bone repair and regeneration. First, the architecture of the two major types of bone (i.e., cortical and trabecular bone) and its relation to the structural integrity of the tissue is discussed. Then, the ultrastructure/composition of bone is explained to provide more detailed information about the physical and chemical makeup of the tissue. Finally, more information is provided about the differences between species, age groups, and genders.

2.2   BONE ARCHITECTURE

The organization of bone is characterized as a highly hierarchical structure as shown in Figure 2.1 [10]. At the macro scale level, human and other mammalian bones are generally classified as cortical or cancellous (trabecular). Cortical bone is found primarily in the shaft of long bones and the outer shell around trabecular bone at the proximal and distal ends of bones and the vertebrae, whereas trabecular bone is located within cortical tissue, in medullary cavities at the ends of long bones, and in the interior of short bones, such as spinal vertebrae [11]. At the sub-microscopic level, lamellae are the basic building blocks of osteons and trabeculae. At the ultrastructural level, a composite of mineral crystals and collagen fibrils build the lamellae.

2.2.1   CORTICAL BONE

Cortical bone comprises about 80% of the total mass of the skeleton. Cortical bone may be classified as primary or secondary bone. Primary bone is tissue laid down on existing bone surfaces during

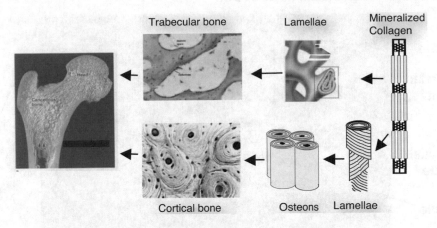

Figure 2.1: Hierarchical structure of bone: collagen fibrils, lamellae, osteons and trabeculae, cortical and trabecular bone.

developmental growth. It may consist of circumferential lamellae, woven tissue, or plexiform tissue. Circumferential lamellar bone consists of lamellae, which run parallel to the bone surface. Located within these circumferential lamellae are primary osteons, which form when blood vessels on the surface of the bone become incorporated into the new periosteal bone. In general, both plexiform and woven bones are found in large and/or fast-growing animals and may be formed at the fracture healing sites. Also, found within cortical bone are void spaces which consist of Haversian canals, Volkmann's canals, and resorption spaces.

Under an optical microscope, most of the microstructural features of human cortical bone can be observed, such as osteons, interstitial tissues, lacunae, and cement lines (Figure 2.2) [12]. Adult human cortical bone contains cylindrical lamellar structures known as osteons or Haversian systems. Osteons vary in size with an average diameter of 200μm. Haversian canals are located in the center of osteons, which contain blood vessels and nerves with an average diameter of $\sim$50μm. The boundary of osteons is the cement line, which separates osteons from the surrounding tissue. Interstitial tissues are those between osteons (Figure 2.2). Depending on the amount of remodeling that has occurred, the interstitial tissue may be a mixture of primary bone or the remnants of primary and secondary osteons. Lacunae are cavities where osteocytes (bone cells buried in the matrix) are located. Osteocytes may communicate with each other through their processes that extend in tiny canals called canaliculi. One of the major microstructural parameters in cortical bone is porosity. Cortical bone is very dense, and its porosity is between 5%-10% [12]. The porosity of cortical bone is mostly contributed by Haversian canals, Volkmann's canals and resorption cavities. Volkmann's canals are transverse canals connecting Haversian canals to each other. Resorption cavities are the temporary spaces created by bone-removing cells (osteoclasts) in the initial stage of bone remodeling. The total volume of lacunae and canaliculi is relatively small, contributing to $\sim$10% of the total porosity [13].

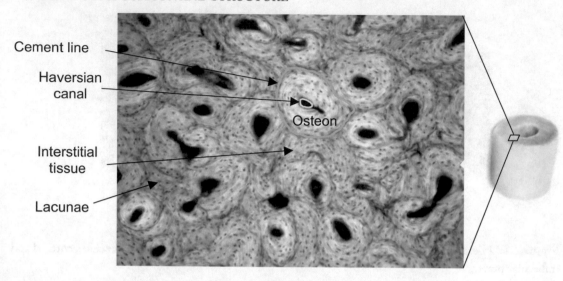

Figure 2.2: Microstructure of cortical bone. Haversian canals are in the center of osteons. Interstitial tissues are the bone lamellae between osteons.

2.2.2   TRABECULAR BONE

Trabecular bone is found in the metaphysis, epiphyses, and medullary cavity of long bones, within flat bones, and within vertebral bodies. It consists of a three-dimensional structure of interconnected plates and rods known as trabeculae, each of which is approximately 200μm thick (Figure 2.3) [12]. Trabecular bone has a porosity between 75%-95%. The pores in trabecular bone are interconnected and filled with bone marrow (Figure 2.3).

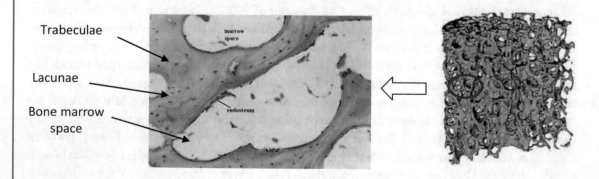

Figure 2.3: Microstructure of trabecular bone: (a) a two-dimensional histology section; (b) a three-dimensional micro computed tomography image.

Microarchitectural parameters have been defined to characterize the morphology of trabecular bone [14,15]. Based on a plate model, the architectural parameters of trabecular bone can be determined from a three-dimensional micro-computed tomography (micro-CT) image, which include bone volume fraction (BV/TV), bone surface-to-volume ratio (BS/BV), trabecular thickness (Tb.Th), trabecular number (Tb.N), trabecular separation (Tb.Sp), and connectivity density (Conn. D). The bone volume fraction (BV/TV) is the ratio of the bone volume to the specimen volume of interest. It has a negative relationship with porosity. The surface-to-volume ratio (BS/BV) characterizes the rate of bone turnover because bone resorption and formation can only occur on bone surfaces. A large surface-to-volume ratio is indicative of high rate of bone turnover. Trabecular thickness (Tb.Th) is defined as the average thickness of trabeculae. Trabecular number (Tb.N) denotes the number of trabeculae in a unit length. Trabecular separation (Tb.Sp) measures the marrow space between trabeculae. Connectivity density (Conn. D) provides a measure of unconnected trabeculae.

2.2.3 DIFFERENCES BETWEEN ANATOMICAL LOCATIONS, AGES, AND GENDERS

2.2.3.1 Anatomical Differences

The bone architecture, especially trabecular bone architecture, is site-dependent [16–19] histomorphometry study has shown that the trabecular BV/TV is much lower in the lumbar spine (8.3%) than that in the femoral neck (15.8%) [16]. In addition, 3D micro-CT investigations indicate that trabecular microstructure has a similar trend in BV/TV at other sites, such as spine, femur, iliac crest, and calcaneus [17]. Eckstein et al. compared architecture of trabecular bone from six different anatomical sites of human bones: distal radius, femoral neck and trochanter, iliac crest, calcaneus, and second lumbar vertebral body [18]. The study reported that the iliac crest displayed the most rod-like trabecular structures, whereas the femoral neck and the calcaneus displayed the most plate-like structures. In addition, the trabeculae are thickest in the femoral neck ($182\pm46\mu$m) and thinnest in the iliac crest ($126\pm19\mu$m).

The architecture of trabecular bone also varies in the different regions at the same anatomic site. For example, there are six regional variations in trabecular architecture at the sagittal section of human lumbar vertebra, showing that the central and antero-superior regions have lower BV/TV and Conn. D compared with that in the posterior-inferior region. In addition, comparing the cranial, mid-vertebra and caudal regions in the ovine lumbar vertebra, Tb.N and Conn. D are significantly higher in the cranial region than the other regions, whereas Tb.Th and Tb.Sp are significantly higher, and Tb.N and Conn.D are significant lower in the mid-vertebra region than the other regions [20].

2.2.3.2 Age Dependence

Beyond middle age (>45 years), the age-related architectural changes in human bone primarily include a decrease in BV/TV and Conn. D for trabecular bone [21], a decrease in the cortical thickness (C.Th) [22], and an increase in the porosity [23,24] for cortical bones. These changes may significantly affect the mechanical behavior of bone, usually leading to the reduced strength and toughness of the tissue.

Trabecular bone undergoes significant age-related changes throughout the skeleton. Perhaps the most significant age-related change is the BV/TV of trabecular bone, which has been found to decrease with age in the femoral head [25], femoral neck [26], lumbar spine [21,25,27], distal forearms [28], iliac crest [29,30], and proximal tibia [31]. In the wrist (a common age-related fracture site), Tb.N and Tb.Th decrease and Tb.Sp increases significantly with age [28]. In the lumbar spine (another common age-related fracture site), trabecular BV/TV, Tb.N and Conn.D decrease significantly with age [27]. Interestingly, age-related changes in tissue anisotropy have been demonstrated in vertebral bodies, with the thickness of the horizontal trabeculae decreasing significantly with age, whereas the trabecular thickness of the vertical trabeculae is age-independent [27]. In addition, age-dependent microstructural changes in the trabecular bone of the tibia metaphysis are reflected in a significant decline of BV/TV, thinning of trabeculae, and a change in microstructure from plate-like to more or less rod-like [31]. Moreover, the mean marrow space volume, bone surface-to-volume ratio also increase with age.

The most significant age-related structural change in cortical bone is porosity. Increases of porosity with aging have been observed in various anatomical locations of skeletal tissues. For instance, the cortex porosity in both the femoral neck and intertrochanter has been shown to significantly increase with age [32,33]. The porosity in femoral diaphyses increases from 4~6% in young adults to over 9% in the elderly [34]. The greater porosity in the elderly is due to the increasing size rather than the number of Haversian canals [35,36]. Age-related increases in the porosity of cortical bone are also demonstrated at other skeletal locations, such as iliac crest [37] and diaphyses of the humerus and tibia [24,38]. In addition to porosity, 3D micro-CT images of femoral diaphyses have shown that the number of Haversian canals is nonlinearly (quadratic) correlated with age in the female population, increases up to ~60 years of age, then start to decrease afterwards [39].

2.2.3.3 Gender Dependence

There are architectural differences between men and women [18,21,40] in which women are more vulnerable to osteoporosis (Figure 2.4). In the forearm radius, the trabecular compartment showed higher BV/TV (except for the mid-shaft region), Tb.N, Conn.D and lower Tb.Sp in men than in women. The cortical compartment showed higher C.Th, tissue area, and moment of inertia in men than in women. However, BV/TV was not gender-dependent in the forearm radius [40]. At the radius and femoral neck, trabecular bone displayed a more plate-like structure, thicker trabeculae, smaller separation, higher connectivity, and a higher degree of anisotropy in men than in women. At the trochanter, men displayed more plate-like structure and thicker trabeculae, but no differences in trabecular separation or other parameters compared with the women. However, at the calcaneus, iliac crest, and second lumbar vertebra none of the bone parameters display significant differences between genders [18].

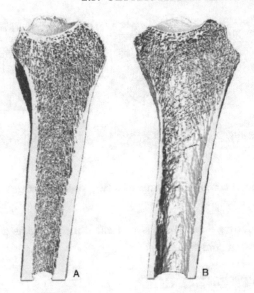

Figure 2.4: HR-pQCT of a healthy and an osteoporotic (distal forearm T-score 3.8) human radius (A) 79 year-old man (B) 79 year-old woman [40]. (Adapted from Bone, 45, Mueller et al., 2009, page 882-891, with permission from Elsevier Inc.)

Women experience more severe changes in bone with increasing age than men although such age-related changes are usually parallel for men and women [18,41]. Men show an age-related increase in bone size (e.g., cross-sectional area of the vertebral bodies), whereas such an increase is not obvious in women. In addition, a higher tendency of disconnection of the trabecular network is present in postmenopausal women than men at a similar age [42]. Women usually have thinner trabeculae in young adulthood and may experience more microstructural damages with increasing age than men. These gender-dependent architectural changes may explain why women are more susceptible to age-related bone fractures than men are.

2.3   ULTRASTRUCTURE/COMPOSITION OF BONE

2.3.1   ULTRASTRUCTURE OF BONE

Similar for both cortical and trabecular bone, organic matrix (mostly type I collagen), apatite mineral (similar to hydroxyapatite crystals) and water together make up the ultrastructure of bone, which can be characterized as a composite of mineral crystals and collagen fibrils. The collagen fibrils are laid down in an organized fashion, and upon mineralization, can form a lamella (Figure 2.5). Therefore, bone material can be simplified as a two-phase composite. Thus, composite materials models can describe its mechanical behavior [43,44]. Bone can be treated as two-phase mixture in two different ways. One is to consider bone tissue as a composite of mineralized collagen fibers and organic matrix

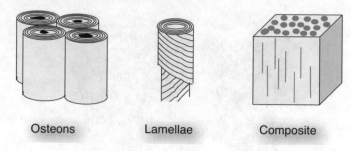

Figure 2.5: Illustration of bone composite model as a two-phase mixture.

from the point view of structure. The other is to treat bone tissue as a composite of mineral and collagen from the point view of composition.

2.3.2   CONSTITUENTS OF BONE

Among the components of bone, the mineral phase occupies up to 60% of the mass or ~40% of the volume of bone. The composition of the mineral phase is mainly calcium (Ca^{2+}) and phosphate (PO_4^{3-}) with a small fraction of carbonates (CO_3^{2-}). In addition, the organic matrix occupies about 40% of the volume of bone [45]. It consists of more than 90% of type I collagen and non-collagenous proteins (e.g., osteocalcin, osteonectin, osteopontin, etc.), which are small in amount, but important in bone structure and bone metabolism. Finally, the water phase occupies up to 25% of the volume of bone.

2.3.2.1   Mineral Phase (Hydroxyapatite)

The X-ray diffraction pattern of bone is similar to that of hydroxyapatite [$Ca_{10}(PO_4)_6(OH)_2$](HA) [75–77]. The crystal structure of HA is shown in Figure 2.6. Mineralization may start with extracellular matrix vesicles or at nucleation sites within the collagen. The former is more associated with rapid bone formation (woven bone), and the latter is more associated with organized bone formation (lamellar bone) [78]. With regard to collagen-based mineralization, crystals are preferentially associated with contiguous gaps in the collagen network [79–81]. The long axis of the plate-like crystal (the crystallographic c-axes) aligns with the long axis of the fibril [81, 82]. One of the unique aspects of bone is that old tissue is continually being replaced with new tissue in a process called bone remodeling. Mineralization in the primary stage occurs rapidly (over a few days) following deposition of osteoid [83] while the secondary stage takes much longer, and its rate of mineral accumulation decreases with time [84]. It may take several months for secondary mineralization to be completed [85]. Recent evidence obtained using Raman spectroscopy indicate that the mineral maturation can occur over two decades [86].

While some X-ray diffraction studies have identified the bone mineral as a hydroxyapatite [75, 77], others have shown that the reflection patterns of bone are different from those of synthetic

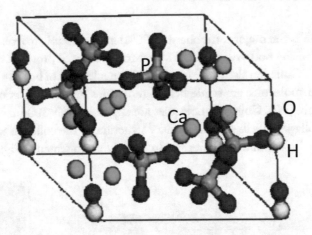

Figure 2.6: Crystal structure of hydroxyapatite.

hydroxyapatite [87]. Initially, the hydroxyl groups (OH-) were not readily identified using Fourier Transform Infrared Spectroscopy (FTIR) or Nuclear Magnetic Resonance (NMR) techniques [88, 89], but recent analysis of bone by NMR spectroscopy indicates that OH- content is about 20% of the content in stoichiometric HA. Nonetheless, some researchers consider it more appropriate to classify the mineral phase as a carbonated apatite ($Ca_5(PO_4CO_3)_3$) [60]. Ultrastructural characteristics that may be used to explain the indistinct diffraction patterns of bone include a non-stoichiometric ratio of calcium to phosphorous, presence of strongly bound water, and deposition of amorphous mineral (tricalcium or octacalcium phosphate) [76].

In addition, electron microscopy and small angle X-ray scattering have shown that bone crystals are quite small (possibly the smallest formed biologically). For example, the plate-like dahllite crystals have length, width, and thickness of 50nm×25nm×1.5∼4.0nm, respectively [90–92]. Given the right concentration and a nucleation agent, mineral (irrespective of composition) will thermodynamically precipitate *in vitro*. In the mammalian skeleton, however, bone mineralization follows collagen organization.

The mineral phase has been considered as the key factor in determining bone mechanical properties. Bone mineral provides bone with stiffness and strength [93, 94]. Additionally, bone mineral crystal orientation determines the anisotropy of bone properties [95]. Bone mineral loss would result in decreases in modulus and strength and increases in fracture risk of bone tissue. Bone mineral content can provide an estimation of some bone mechanical properties and has served as one risk factor used to predict bone fractures [94, 96–99].

2.3.2.2 Collagen

Type I collagen protein is the major component (>90%) in the organic matrix. There are many types of collagen identified in the body. As a fibrous structural protein, it provides the bone with strength and flexibility [46–52]. Collagen also provides spaces for nucleation of bone mineral crystals [53,54].

Type I collagen molecules are triple helical molecules which consist of two $\alpha 1$ polypeptide chains and one $\alpha 2$ chain [60]. Collagen molecules are approximately 300nm in length and 1.2nm in diameter [61]. The collagen fibrils are a secondary structure of the collagen network, with collagen molecules being cross-linked and staggered in an orderly arrangement (Figure 2.7). The collagen

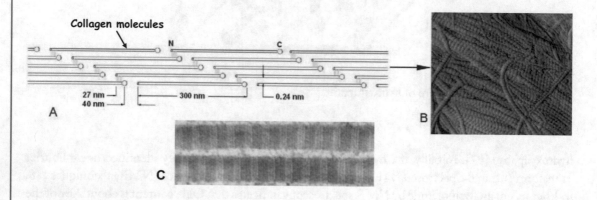

Figure 2.7: Collagen molecules arrange into fibrils in a staggered fashion with cross-links connecting the C-terminal to the N-terminal of neighboring molecule (A). The overlap regions of molecules create a banding pattern visible with atomic force microscopy (B) or with transmission electron microscopy (C).

cross-links are a salient feature of bone because they not only organize fibrillation, but they also contribute to the mineralization process [53,55], thus affecting the mechanical behavior of bone [56–59] A structural model that represents how collagen molecules are assembled into collagen fibrils has been proposed by Hodge et al. [62]. As shown in Figure 2.7, adjacent collagen molecules in the horizontal direction are separated by a distance of 35nm to 40nm [62]. The gaps of vertically aligned molecules are offset by a distance of 64nm to 70nm [63]. The resulting overlap (27nm) leads to the banding pattern between collagen fibrils, which can be observed with electron or atomic force microscopy (Figure 2.7). Neighboring collagen molecules are interconnected through two types of enzymatic cross-links. One is immature enzymatic cross-links, such as hydroxylysinonorleucine (HLNL) and dihydroxylysinonorleucine (DHLNL) [64]. The other type is mature enzymatic cross-links, such as hydroxylysyl-pyridinoline (HP) and lysyl-pyridinoline (LP) [64]. Enzymatic cross-links are formed through the lysyl and hydroxylysyl residues, which are available at both C- and N-terminal ends [64,65].

Collagen orientation affects bone mechanical properties [66–69]. However, denaturing of the collagen matrix does not have profound effects on bone strength [70–72] and viscoelastic properties [74]. Denaturing collagen does cause a significant decrease in the toughness of bone, but it has little effect on the stiffness of bone [73]. Collagen integrity is an important determinant of fracture risk [71,72].

2.3.2.3 Non-Collagenous Proteins (NCPs)

In addition to collagen, the organic matrix has other non-collagenous components, including small, but important amounts of non-collagenous proteins such as osteocalcin , osteonectin, osteopontin, etc. The structural functions of noncollagenous proteins are not very clear. These proteins may influence the events associated with bone remodeling, such as recruitment, attachment, differentiation, and activity of bone cells [108,109]. Among the proteins, the most abundant one is osteocalcin, which is produced by osteoblasts and believed to be connected with bone calcification [6,110,111]. A detailed list of non-collagenous proteins and their possible functions is shown in Table 2.1 [112,113]. As structural proteins, non-collagenous proteins also contribute to bone mechanical integrity including strength, hardness and flexibility [114–116].

Table 2.1: Major non-collagens proteins in bone and possible functions [112,113].

Non-collagen Proteins	Possible functions
Alkaline Phosphatase	• A phosphotransferase • Potential calcium carrier • Hydrolizes inhibitors of mineral deposition
Alpha 2 HS-Glycoprotein	• Participate in the development of the tissues • Possible mineralization inhibitor
Biglycan	• Play a role in the mineralization of bone
Bone Sialoprotein (BSP)	• May initiate mineralization
Decorin	• May regulate fibril diameter • May initiate cell attachment to fribonectin
Matrix Gla Protein	• May initiate mineralization
Osteocalcin	• May regulate activity of osteoclasts and their precursors • May mark tuning point between bone resorption and formation • Regulate mineral maturation
Osteonectin	• May mediate deposition of hydroxyapatite • Binds to growth fractors
Osteopontin	• May regulate cell attachment

2.3.2.4 Water

Water is distributed throughout bone in three major forms: freely mobile in vascular-lacunar-canalicular space, bound to the surface of collagen network and mineral phase, and as hydroxyl component of other molecules [45,100]. For example, free water residing in the marrow-vascularosteocytic space of canine cortical bone may take up 13.74% to 22.84% of volume. The bound water, on the other hand, occupies an additional ~6.97% of the volume of the tissue [101].

H_2O is by nature a polar molecule and associates itself naturally with mineral (PO_4^{3-} or Ca^{2+}) and collagen (glycine, hydroxyproline, carboxyl, and hydroxylysine). The studies on the hydration of collagenous tissue (human dura mater and rat-tail tendons) with dynamic mechanical spectroscopy indicate that water does associate with collagen at two levels: hydrogen bonding occurs on and between the hydroxyl group of hydroxyproline and the polar side chains of collagen molecules, respectively [102, 103].

Water, together with mineral and collagen, plays an important role in bone mechanical properties [104], as evidenced by an inverse correlation between the water content and bone mechanical properties [105]. This has been supported by several other studies. For instance, Fosse et al. showed that low water content significantly increased the stiffness of morsellized bone, while the high water content significantly reduced it [106]. Nyman et al. systematically studied the functions of mobile and bound water [107]. They found that bound water was associated with both strength and failure energy dissipation in bone, whereas mobile water was correlated with modulus of elasticity and the porosity of bone. Moreover, water plays a significant role in the viscoelastic behavior of bone [74], showing that the relaxation rate for the Debye relaxation decreases linearly with water content. This linear relation is attributed to the collagen rearrangement by the varying water content and nucleation of microcracks [74].

2.3.3 DIFFERENCES BETWEEN ANATOMICAL LOCATIONS, AGES, AND GENDERS

The composition of bone varies with the anatomical locations, ages, and genders. Looker et al. studied bone mineral density (BMD) at skeletal sub-regions of pelvis, lumber spine, right leg and left arm for both men and women [117]. The number of samples is more than 6,500 for each gender. For both genders, BMD is highest at the pelvis and lowest at the left arm. For all the sub-regions, BMD is higher in men than in women. The study also demonstrated that BMD keeps decreasing with aging at all the sub-regions in women and at sub-regions of pelvis, right leg and left arm in men. Similar BMD changes were found in other studies [118–120]. Decrease of BMD with aging was not observed at lumber spine in men [117].

The change of collagen fiber orientation is both age- and gender-related [121, 122]. Overall, collagen fibrils become more transversely oriented in elderly. Goldman observed that the proportion of transverse collagen fibers decreased between the young and middle groups, followed by a later increase between the middle and older groups. Although a similar trend exists between men and women, women demonstrated higher variability among groups than men [121].

The stability of collagen network is determined by both collagen molecules and intermolecular cross-links. There is an age-related decline in the intrinsic collagen content. However, no evidence indicates biochemical modifications of collagen during aging for trabecular bone (bone biopsies of iliac crest) [46]. A study on human femur cortical bone by Wang et al. demonstrates that denaturation of the noncalcified collagenous matrix in bone increases with increasing age. In addition, such collagen denaturation in osteoid may correlate with nonenzymatic collagen cross-links as well as the strength and toughness of bone [123]. Eyre et al. found that the enzymatic immature cross-links consistently decrease whereas the mature cross-links increase until around the age of 28 years, and then slightly decrease after the age of 50 [124]. Nyman et al. observed the enzymatic mature cross-links in women decrease after middle age [58]. They also demonstrated an increase in pentosidine concentration (a marker for non-enzymatic glycation induced collagen cross-links) with age in bulk tissue and in the secondary osteonal and interstitial tissues [125–127]. Odetti et al. found that the pentosidine concentration in cortical bone increased exponentially with age [128]. However, pentosidine concentrations in trabecular bone have not exhibited an increase with age and have demonstrated high variability [128, 129].

2.4   DIFFERENCES BETWEEN SPECIES

Differences in the composition and structure between species are the most important factors that need to be discussed when considering the scaffold design for a chosen animal model. The most often used animal models in bone studies include the rat, mouse, dog, cow, pig, sheep, rabbit, etc. An ideal model that can perfectly mimic human bone does not exist. By examining the relative differences between the femoral and lumbar spine trabecular bone samples from the human, dog, cow, pig, sheep and rat, the highest similarity is demonstrated between human and canine bones, and the lowest similarity resides between human and rat bones [130].

2.4.1   DIFFERENCES IN TISSUE ARCHITECTURE

Bone structural differences between species are more important at micro scale levels. Modern imaging techniques have enabled observation and quantification of bone microstructure, such as light microscopy [131] and micro-CT [132, 133]. The structural differences between species are more obvious in cortical bone than in trabecular bone. The cortical bone of the human femur consists of secondary osteons and interstitial tissue, whereas the basic structure of bone from the femur of the pig, cow, and sheep is the primary vascular plexiform bone. Irregular Haversian bone tissues are sparsely found at the periosteal border and the posterior aspect of the femur from sheep and cows. The femoral diaphysis of rabbit is mainly composed of primary vascular woven bone tissue. The microscopic structure of femoral diaphysis of rat is comprised of nonvascular woven bone tissue, and no secondary osteons could be found [134]. On the other hand, the trabecular bone architecture of all these species is similar. The differences that can be characterized are reflected in some parameters of trabecular structure, such as trabecular path-length, cavity length, bone volume fraction, etc. For example, mean cavity path lengths typically is $1,200\mu m$ for the adult man, and $350\mu m$ in the miniature pig; mean trabecular path lengths typically $220\mu m$ for the adult man, and $280\mu m$ for the miniature pig; percentage bone volume is around 16% for adult man and around 45% for the miniature pig [131].

Table 2.2: Compositional differences of bone between species [130].

Property	Human	Dog	Pig	Cow	Chicken	Sheep	Rat
Ash (%)	61.0~64.2	60.8~64.4	63.4~67.9	61.3~65.8	60.8~69.7	62.6~69.7	62.1~67.6
Hydroxyproline (μg/mg dry)	28.7~30.7	26.2~30.8	20.2~23.4	26.7~32.8	23.6~27.3	19.8~27.9	18.2~23.7
Total protein (μg/mg dry)	40.1~77.5	57.9~77.5	75.6~124.4	49.7~79.7	52.1~150.8	68.1~126.1	58.6~101

2.4.2   DIFFERENCES IN TISSUE COMPOSITION/ULTRASTRUCTURE

Compositional/ultrastructural differences of bone exist among different species. For example, bone samples from the femoral cortex and lumber spine of human, dog, cow, pig, sheep, rat and chicken have demonstrated marked distinctions [130]. The ash content of cortical bone samples from all of the species is very similar, whereas the amount of the organic phase is significantly different between the species (Table 2.2). Comparing the compositional properties between the species, only dog bone has the most similar composition with that of human bone. All the compositional parameters of trabecular bone derived from lumbar spine followed the similar trend as of cortical bone for all of the species, but with less variation. Only the trabecular samples derived from cows showed a higher-than-expected amount of extractable proteins compared with the results from cortical bone samples of other species.

2.5   SUMMARY

Bone can be classified as cortical or trabecular. Both types of bone are composed of collagen, mineral, water and non-collagenous proteins. The amount, the morphology, the quality and the arrangement of these bone constituents determine the mechanical and biochemical behaviors of bone. There are compositional and structural differences among species (such as human and rat), anatomical sites (such as lumbar spine and femur head; cortical and trabecular bone), age groups (such as middle-aged and elderly) and genders. Understanding the composition/structure of natural bone tissues is useful to (1) the fabrication of tissue engineered bone products that mimic the natural tissues; (2) the development of a functional design for the architecture of scaffolds; and (3) the selection of adequate animal models for tissue engineering research and pre-clinical trials of tissue engineering products.

REFERENCES

[1] Langer, R., Vacanti, J.P., Tissue engineering. Science, 1993. **260**(5110). p. 920–6. 2.1

[2] Donati, D., Zolezzi, C., Tomba, P., Vigano, A., Bone grafting: historical and conceptual review, starting with an old manuscript by Vittorio Putti. Acta Orthop, 2007. **78**(1). p. 19–25. DOI: 10.1080/17453670610013376 2.1

[3] Schmidt, C., Ignatius, A.A., Claes, L.E., Proliferation and differentiation parameters of human osteoblasts on titanium and steel surfaces. J Biomed Mater Res, 2001. **54**(2). p. 209–15. DOI: 10.1002/1097-4636(200102)54:2%3C209::AID-JBM7%3E3.0.CO;2-7 2.1

[4] Thompson, D.D., Simmons, H.A., Pirie, C.M., Ke, H.Z., FDA Guidelines and animal models for osteoporosis. Bone, 1995. **17**(4 Suppl). p. 125S-133S. DOI: 10.1016/8756-3282(95)00285-L 2.1

[5] Hunziker, E.B., Tissue engineering of bone and cartilage. From the preclinical model to the patient. Novartis Found Symp, 2003. **249**. p. 70–8; discussion 78–85, 170-4, 239-41. DOI: 10.1002/0470867973.ch6 2.1

[6] Mbuyi-Muamba, J.M., Dequeker, J., Gevers, G., Collagen and non-collagenous proteins in different mineralization stages of human femur. Acta Anat (Basel), 1989. **134**(4). p. 265–8. DOI: 10.1159/000146700 2.1, 2.3.2.3

[7] Puelacher, W.C., Vacanti, J.P., Ferraro, N.F., Schloo, B., Vacanti, C.A., Femoral shaft reconstruction using tissue-engineered growth of bone. Int J Oral Maxillofac Surg, 1996. **25**(3). p. 223–8. DOI: 10.1016/S0901-5027(96)80035-X 2.1

[8] Khan, S.N.,Lane, J.M., Spinal fusion surgery: animal models for tissue-engineered bone constructs. Biomaterials, 2004. **25**(9). p. 1475–85. DOI: 10.1016/S0142-9612(03)00491-5 2.1

[9] Liebschner, M.A., Biomechanical considerations of animal models used in tissue engineering of bone. Biomaterials, 2004. **25**(9). p. 1697–714. DOI: 10.1016/S0142-9612(03)00515-5 2.1

[10] Rho, J.Y., Kuhn-Spearing, L., Zioupos, P., Mechanical properties and the hierarchical structure of bone. Medical Engineering & Physics, 1998. **20**(2). p. 92–102. DOI: 10.1016/S1350-4533(98)00007-1 2.2

[11] Martin, R.B.,Burr, D.B., The structure, function, and adaptation of compact bone. 1989, Raven Press. 2.2

[12] Martin, R.B., Burr, D.B., Sharkey, N.A., Skeletal Tissue Mechanics. 1998, Springer. 2.2.1, 2.2.1, 2.2.2

[13] Schaffler, M.B.,Burr, D.B., Stiffness of compact bone: effects of porosity and density. J Biomech, 1988. **21**(1). p. 13–6. DOI: 10.1016/0021-9290(88)90186-8 2.2.1

[14] Feldkamp, L.A., Goldstein, S.A., Parfitt, A.M., Jesion, G., Kleerekoper, M., The direct examination of three-dimensional bone architecture in vitro by computed tomography. J Bone Miner Res, 1989. **4**(1). p. 3–11. 2.2.2

[15] Goulet, R.W., Goldstein, S.A., Ciarelli, M.J., Kuhn, J.L., Brown, M.B., Feldkamp, L.A., The relationship between the structural and orthogonal compressive properties of trabecular bone. J Biomech, 1994. **27**(4). p. 375–89. DOI: 10.1016/0021-9290(94)90014-0 2.2.2

[16] Amling, M., Herden, S., Posl, M., Hahn, M., Ritzel, H., Delling, G., Heterogeneity of the skeleton: comparison of the trabecular microarchitecture of the spine, the iliac crest, the femur, and the calcaneus. J Bone Miner Res, 1996. **11**(1). p. 36–45. 2.2.3.1

[17] Hildebrand, T., Laib, A., Muller, R., Dequeker, J., Ruegsegger, P., Direct three-dimensional morphometric analysis of human cancellous bone: microstructural data from spine, femur, iliac crest, and calcaneus. J Bone Miner Res, 1999. **14**(7). p. 1167–74. DOI: 10.1359/jbmr.1999.14.7.1167 2.2.3.1

[18] Eckstein, F., Matsuura, M., Kuhn, V., Priemel, M., Mueller, R., Link, T.M., Lochmueller, E.M., Sex differences of human trabecular bone microstructure in aging are site-dependent. Journal of Bone and Mineral Research, 2007. **22**(6). p. 817–824. DOI: 10.1359/jbmr.070301 2.2.3.1, 2.2.3.3, 2.2.3.3

[19] Majumdar, S., Kothari, M., Augat, P., Newitt, D.C., Link, T.M., Lin, J.C., Lang, T., Lu, Y., Genant, H.K., High-resolution magnetic resonance imaging: Three-dimensional tra-becular bone architecture and biomechanical properties. Bone, 1998. **22**(5). p. 445–454. DOI: 10.1016/S8756-3282(98)00030-1 2.2.3.1

[20] Kennedy, O.D., Brennan, O., Rackard, S.M., O'Brien, F.J., Taylor, D., Lee, T.C., Vari-ation of trabecular microarchitectural parameters in cranial, caudal and mid-vertebral regions of the ovine L3 vertebra. Journal of Anatomy, 2009. **214**(5). p. 729–735. DOI: 10.1111/j.1469-7580.2009.01054.x 2.2.3.1

[21] Chen, H., Shoumura, S., Emura, S., Bunai, Y., Regional variations of vertebral trabecular bone microstructure with age and gender. Osteoporosis International, 2008. **19**(10). p. 1473–1483. DOI: 10.1007/s00198-008-0593-3 2.2.3.2, 2.2.3.3

[22] Ritzel, H., Amling, M., Posl, M., Hahn, M., Delling, G., The thickness of human vertebral cortical bone and its changes in aging and osteoporosis: A histomorphometric analysis of the complete spinal column from thirty-seven autopsy specimens. Journal of Bone and Mineral Research, 1997. **12**(1). p. 89–95. DOI: 10.1359/jbmr.1997.12.1.89 2.2.3.2

[23] Brockstedt, H., Bollerslev, J., Melsen, F., Mosekilde, L., Cortical bone remodeling in autoso-mal dominant osteopetrosis: A study of two different phenotypes. Bone, 1996. **18**(1). p. 67–72. DOI: 10.1016/8756-3282(95)00424-6 2.2.3.2

[24] NyssenBehets, C., Duchesne, P.Y., Dhem, A., Structural changes with aging in cortical bone of the human tibia. Gerontology, 1997. **43**(6). p. 316–325. DOI: 10.1159/000213871 2.2.3.2

[25] Stauber, M., Muller, R., Age-related changes in trabecular bone microstructures: global and local morphometry. Osteoporosis International, 2006. **17**(4). p. 616–626. DOI: 10.1007/s00198-005-0025-6 2.2.3.2

[26] Bell, K.L., Garrahan, N., Kneissel, M., Loveridge, N., Grau, E., Stanton, M., Reeve, J., Corti-cal and cancellous bone in the human femoral neck: Evaluation of an interactive image analysis system. Bone, 1996. **19**(5). p. 541–548. DOI: 10.1016/S8756-3282(96)00245-1 2.2.3.2

[27] Thomsen, J.S., Ebbesen, E.N., Mosekilde, L., Age-related differences between thinning of horizontal and vertical trabeculae in human lumbar bone as assessed by a new computerized method. Bone, 2002. **31**(1). p. 136–142. DOI: 10.1016/S8756-3282(02)00801-3 2.2.3.2

[28] Khosla, S., Riggs, B.L., Atkinson, E.J., Oberg, A.L., McDaniel, L.J., Holets, M., Peterson, J.M., Melton, L.J., Effects of sex and age on bone microstructure at the ultradistal radius: A population-based noninvasive in vivo assessment. Journal of Bone and Mineral Research, 2006. **21**(1). p. 124–131. DOI: 10.1359/JBMR.050916 2.2.3.2

[29] Thomsen, J.S., Barlach, J., Mosekilde, L., Determination of connectivity density in human iliac crest bone biopsies assessed by a computerized method. Bone, 1996. **18**(5). p. 459–465. DOI: 10.1016/8756-3282(96)00048-8 2.2.3.2

[30] Thomsen, J.S., Ebbesen, E.N., Mosekilde, L., Relationships between static histomorphometry and bone strength measurements in human iliac crest bone biopsies. Bone, 1998. **22**(2). p. 153–163. DOI: 10.1016/S8756-3282(97)00235-4 2.2.3.2

[31] Ding, M., Odgaard, A., Linde, F., Hvid, I., Age-related variations in the microstructure of human tibial cancellous bone. Journal of Orthopaedic Research, 2002. **20**(3). p. 615–621. DOI: 10.1016/S0736-0266(01)00132-2 2.2.3.2

[32] Vajda, E.G., Bloebaum, R.D., Age-related hypermineralization in the female proximal human femur. Anat Rec, 1999. **255**(2). p. 202–11. DOI: 10.1002/(SICI)1097-0185(19990601)255:2%3C202::AID-AR10%3E3.0.CO;2-0 2.2.3.2

[33] Bousson, V., Peyrin, F., Bergot, C., Hausard, M., Sautet, A., Laredo, J.D., Cortical bone in the human femoral neck: three-dimensional appearance and porosity using synchrotron radiation. J Bone Miner Res, 2004. **19**(5). p. 794–801. DOI: 10.1359/JBMR.040124 2.2.3.2

[34] Feik, S.A., Thomas, C.D., Clement, J.G., Age-related changes in cortical porosity of the midshaft of the human femur. J Anat, 1997. **191** ((Pt 3)). p. 407–416. DOI: 10.1046/j.1469-7580.1997.19130407.x 2.2.3.2

[35] Stein, M.S., Feik, S.A., Thomas, C.D., Clement, J.G., Wark, J.D., An automated analysis of intracortical porosity in human femoral bone across age. J Bone Miner Res, 1999. **14**(4). p. 624–32. DOI: 10.1359/jbmr.1999.14.4.624 2.2.3.2

[36] Thomas, C.D.L., Feik, S.A., Clement, J.G., Increase in pore area, and not pore density, is the main determinant in the development of porosity in human cortical bone. Journal of Anatomy, 2006. **209**(2). p. 219–230. DOI: 10.1111/j.1469-7580.2006.00589.x 2.2.3.2

[37] Brockstedt, H., Kassem, M., Eriksen, E.F., Mosekilde, L., Melsen, F., Age-Related and Sex-Related Changes in Iliac Cortical Bone Mass and Remodeling. Bone, 1993. **14**(4). p. 681–691. DOI: 10.1016/8756-3282(93)90092-O 2.2.3.2

[38] Lavaljeantet, A.M., Bergot, C., Carroll, R., Garciaschaefer, F., Cortical Bone Senescence and Mineral Bone-Density of the Humerus. Calcified Tissue International, 1983. **35**(3). p. 268–272. DOI: 10.1007/BF02405044 2.2.3.2

[39] Cooper, D.M.L., Thomas, C.D.L., Clement, J.G., Turinsky, A.L., Sensen, C.W., Hallgrimsson, B., Age-dependent change in the 3D structure of cortical porosity at the human femoral midshaft. Bone, 2007. **40**(4). p. 957–965. DOI: 10.1016/j.bone.2006.11.011 2.2.3.2

[40] Mueller, T.L., van Lenthe, G.H., Stauber, M., Gratzke, C., Eckstein, F., Muller, R., Regional, age and gender differences in architectural measures of bone quality and their correlation to bone mechanical competence in the human radius of an elderly population. Bone, 2009. **45**(5). p. 882–91. DOI: 10.1016/j.bone.2009.06.031 2.2.3.3, 2.2.3.3, 2.4

[41] Lochmuller, E.M., Matsuura, M., Bauer, J., Hitzl, W., Link, T.M., Muller, R., Eckstein, F., Site-Specific Deterioration of Trabecular Bone Architecture in Men and Women With Advancing Age. Journal of Bone and Mineral Research, 2008. **23**(12). p. 1964–1973. DOI: 10.1359/jbmr.080709 2.2.3.3

[42] Mosekilde, L., Age-related changes in bone mass, structure, and strength - effects of loading. Zeitschrift Fur Rheumatologie, 2000. **59**. p. 1–9. DOI: 10.1007/s003930070031 2.2.3.3

[43] Currey, J.D., Strength of Bone. Nature, 1962. **195**(4840). p. 513-&. 2.3.1

[44] Bonfield, W., Li, C.H., Anisotropy of Nonelastic Flow in Bone. Journal of Applied Physics, 1967. **38**(6). p. 2450-&. DOI: 10.1063/1.1709926 2.3.1

[45] Elliott, S.R., Robinson, R.A., The water content of bone. I. The mass of water, inorganic crystals, organic matrix, and CO_2 space components in a unit volume of the dog bone. J Bone Joint Surg Am, 1957. **39-A**(1). p. 167–88. 2.3.2, 2.3.2.4

[46] Bailey, A.J., Sims, T.J., Ebbesen, E.N., Mansell, J.P., Thomsen, J.S., Mosekilde, L., Age-related changes in the biochemical properties of human cancellous bone collagen: relationship to bone strength. Calcif Tissue Int, 1999. **65**(3). p. 203–10. DOI: 10.1007/s002239900683 2.3.2.2, 2.3.3

[47] Boskey, A.L., Wright, T.M., Blank, R.D., Collagen and bone strength. J Bone Miner Res, 1999. **14**(3). p. 330–5. DOI: 10.1359/jbmr.1999.14.3.330 2.3.2.2

[48] Lange, M., Qvortrup, K., Svendsen, O.L., Flyvbjerg, A., Nowak, J., Petersen, M.M., K, O.L., Feldt-Rasmussen, U., Abnormal bone collagen morphology and decreased bone strength in growth hormone-deficient rats. Bone, 2004. **35**(1). p. 178–85. DOI: 10.1016/j.bone.2004.02.025 2.3.2.2

[49] Martin, R.B.,Ishida, J., The relative effects of collagen fiber orientation, porosity, density, and mineralization on bone strength. J Biomech, 1989. **22**(5). p. 419–26. DOI: 10.1016/0021-9290(89)90202-9 2.3.2.2

[50] Viguet-Carrin, S., Garnero, P., Delmas, P.D., The role of collagen in bone strength. Osteoporos Int, 2006. **17**(3). p. 319–36. DOI: 10.1007/s00198-005-2035-9 2.3.2.2

[51] Wang, X., Bank, R.A., TeKoppele, J.M., Hubbard, G.B., Athanasiou, K.A., Agrawal, C.M., Effect of collagen denaturation on the toughness of bone. Clin Orthop Relat Res, 2000(371). p. 228–39. 2.3.2.2

[52] Wang, X., Shen, X., Li, X., Agrawal, C.M., Age-related changes in the collagen network and toughness of bone. Bone, 2002. **31**(1). p. 1–7. DOI: 10.1016/S8756-3282(01)00697-4 2.3.2.2

[53] Willems, N.M.B.K., Bank, R.A., Mulder, L., Langenbach, G.E.J., Grunheid, T., Zentner, A., van Eijden, T.M.G.J., Mineralization Is Related to Collagen Cross-links in Growing Cancellous and Cortical Bone. Journal of Bone and Mineral Research, 2008. **23**. p. S155-S155. 2.3.2.2, 2.3.2.2

[54] Strates, B.S.,Nimni, M.E., Mineralization and bone induction in crosslinked collagen and demineralized bone. Journal of Dental Research, 1997. **76**. p. 1629–1629. 2.3.2.2

[55] Saito, M., Fujii, K., Marumo, K., Degree of mineralization-related collagen crosslinking in the femoral neck cancellous bone in cases of hip fracture and controls. Calcified Tissue International, 2006. **79**(3). p. 160–168. DOI: 10.1007/s00223-006-0035-1 2.3.2.2

[56] Oxlund, H., Barckman, M., Ortoft, G., Andreassen, T.T., Reduced concentrations of collagen cross-links are associated with reduced strength of bone. Bone, 1995. **17**(4 Suppl). p. 365S-371S. DOI: 10.1016/8756-3282(95)00328-B 2.3.2.2

[57] Fritsch, A., Hellmich, C., Dormieux, L., Ductile sliding between mineral crystals followed by rupture of collagen crosslinks: experimentally supported micromechanical explanation of bone strength. J Theor Biol, 2009. **260**(2). p. 230–52. DOI: 10.1016/j.jtbi.2009.05.021 2.3.2.2

[58] Nyman, J.S., Roy, A., Acuna, R.L., Gayle, H.J., Reyes, M.J., Tyler, J.H., Dean, D.D., Wang, X., Age-related effect on the concentration of collagen crosslinks in human osteonal and interstitial bone tissue. Bone, 2006. **39**(6). p. 1210–7. DOI: 10.1016/j.bone.2006.06.026 2.3.2.2, 2.3.3

[59] Tierney, C.M., Haugh, M.G., Liedl, J., Mulcahy, F., Hayes, B., O'Brien, F.J., The effects of collagen concentration and crosslink density on the biological, structural and mechanical properties of collagen-GAG scaffolds for bone tissue engineering. J Mech Behav Biomed Mater, 2009. **2**(2). p. 202–9. DOI: 10.1016/j.jmbbm.2008.08.007 2.3.2.2

[60] Weiner, S., Traub, W., Bone-Structure - from Angstroms to Microns. Faseb Journal, 1992. **6**(3). p. 879–885. 2.3.2.1, 2.3.2.2

[61] Weiner, S., Traub, W., Wagner, H.D., Lamellar bone: Structure-function relations. Journal of Structural Biology, 1999. **126**(3). p. 241–255. DOI: 10.1006/jsbi.1999.4107 2.3.2.2

[62] Ramachandran, G.N., Aspects of protein structure, Academic Press, p. 299–300. 1963 2.3.2.2

[63] Hodge, A.J., Molecular-Models Illustrating the Possible Distributions of Holes in Simple Systematically Staggered Arrays of Type-I Collagen Molecules in Native-Type Fibrils. Connective Tissue Research, 1989. **21**(1–4). p. 467-477. DOI: 10.3109/03008208909050004 2.3.2.2

[64] Bailey, A.J., Paul, R.G., Knott, L., Mechanisms of maturation and ageing of collagen. Mechanisms of Ageing and Development, 1998. **106**(1–2). p. 1-56. DOI: 10.1016/S0047-6374(98)00119-5 2.3.2.2

[65] Anderson, J.J. and Garnero, S.C., Calcium and Phosphorus in Health and Disease, CRC Press: Boca Raton. p. 127–145. 1996 2.3.2.2

[66] Evans, F.G., Vincente.R, Relation of Collagen Fiber Orientation to Some Mechanical Properties of Human Cortical Bone. Journal of Biomechanics, 1969. **2**(1). p. 63-&. DOI: 10.1016/0021-9290(69)90042-6 2.3.2.2

[67] Vincente.R, Relation of Collagen Fiber Orientation to Some Mechanical Properties of Human Cortical Bone. Anatomical Record, 1968. **160**(2). DOI: 10.1016/0021-9290(69)90042-6 2.3.2.2

[68] Vincentelli, R., Evans, F.G., Relations among mechanical properties, collagen fibers, and calcification in adult human cortical bone. Journal of Biomechanics, 1971. **4**(3). p. 193–201. DOI: 10.1016/0021-9290(71)90004-2 2.3.2.2

[69] Martin, R.B., Lau, S.T., Mathews, P.V., Gibson, V.A., Stover, S.M., Collagen fiber organization is related to mechanical properties and remodeling in equine bone. A comparison of two methods. Journal of Biomechanics, 1996. **29**(12). p. 1515–1521. DOI: 10.1016/S0021-9290(96)80002-9 2.3.2.2

[70] Jepsen, K.J., Goldstein, S.A., Kuhn, J.L., Schaffler, M.B., Bonadio, J., Type-I collagen mutation compromises the post-yield behavior of Mov13 long bone. J Orthop Res, 1996. **14**(3). p. 493–9. DOI: 10.1002/jor.1100140320 2.3.2.2

[71] Wang, X., Li, X., Bank, R.A., Agrawal, C.M., Effects of collagen unwinding and cleavage on the mechanical integrity of the collagen network in bone. Calcif Tissue Int, 2002. **71**(2). p. 186–92. DOI: 10.1007/s00223-001-1082-2 2.3.2.2

[72] Wynnyckyj, C., Omelon, S., Savage, K., Damani, M., Chachra, D., Grynpas, M.D., A new tool to assess the mechanical properties of bone due to collagen degradation. Bone, 2009. **44**(5). p. 840–848. DOI: 10.1016/j.bone.2008.12.014 2.3.2.2

[73] Wang, X., Bank, R.A., TeKoppele, J.M., Mauli Agrawal, C., The role of collagen in determining bone mechanical properties. Journal of Orthopaedic Research, 2001. **19**(6). p. 1021–1026. DOI: 10.1016/S0736-0266(01)00047-X 2.3.2.2

[74] Sasaki, N., Enyo, A., Viscoelastic properties of bone as a function of water content. J Biomech, 1995. **28**(7). p. 809–15. DOI: 10.1016/0021-9290(94)00130-V 2.3.2.2, 2.3.2.4

[75] Roseberry, H.H., Hastings, A.B., Morse, J.K., X- ray analysis of bone and teeth. Journal of Biological Chemistry, 1931. **90**(2). p. 395–U2. 2.3.2.1, 2.3.2.1

[76] Glimcher, M.J., The nature of the mineral component of bone and the mechanism of calcification. Instr Course Lect, 1987. **36**. p. 49–69. 2.3.2.1, 2.3.2.1

[77] Posner, A.S., Crystal Chemistry of Bone Mineral. Physiological Reviews, 1969. **49**(4). p. 760-&. 2.3.2.1, 2.3.2.1

[78] Lowenstam, H.A., Weiner, S., On Biomineralization. 1989, Oxford University Press, Inc. 2.3.2.1

[79] Landis, W.J., Hodgens, K.J., Arena, J., Song, M.J., McEwen, B.F., Structural relations between collagen and mineral in bone as determined by high voltage electron microscopic tomography. Microscopy Research and Technique, 1996. **33**(2). p. 192–202. DOI: 10.1002/(SICI)1097-0029(19960201)33:2%3C192::AID-JEMT9%3E3.0.CO;2-V 2.3.2.1

[80] Robinson, R.A., Watson, M.L., Collagen-Crystal Relationships in Bone as Seen in the Electron Microscope. Anatomical Record, 1952. **114**(3). DOI: 10.1002/ar.1091140302 2.3.2.1

[81] Weiner, S., Traub, W., Organization of Hydroxyapatite Crystals within Collagen Fibrils. Febs Letters, 1986. **206**(2). p. 262–266. DOI: 10.1016/0014-5793(86)80993-0 2.3.2.1

[82] Landis, W.J., Song, M.J., Leith, A., Mcewen, L., Mcewen, B.F., Mineral and Organic Matrix Interaction in Normally Calcifying Tendon Visualized in 3 Dimensions by High-Voltage Electron-Microscopic Tomography and Graphic Image-Reconstruction. Journal of Structural Biology, 1993. **110**(1). p. 39–54. DOI: 10.1006/jsbi.1993.1003 2.3.2.1

[83] Wergedal, J.E., Baylink, D.J., Electron-Microprobe Measurements of Bone Mineralization Rate in-Vivo. American Journal of Physiology, 1974. **226**(2). p. 345–352. 2.3.2.1

[84] Marotti, G., Zallone, A.Z., Favia, A., Quantitative Analysis on Rate of Secondary Bone Mineralization. Calcified Tissue Research, 1972. **10**(1). p. 67-&. DOI: 10.1007/BF02012537 2.3.2.1

[85] Amprino, R., Investigations on some physical properties of bone tissue. Acta Anat (Basel), 1958. **34**. p. 161–186. DOI: 10.1159/000141381 2.3.2.1

[86] Akkus, O., Polyakova-Akkus, A., Adar, F., Schaffler, M.B., Aging of microstructural compartments in human compact bone. J Bone Miner Res, 2003. **18**(6). p. 1012–9. DOI: 10.1359/jbmr.2003.18.6.1012 2.3.2.1

[87] Boskey, A., Bone mineral crystal size. Osteoporosis International, 2003. **14**. p. S16-S20. 2.3.2.1

[88] Rey, C., Miquel, J.L., Facchini, L., Legrand, A.P., Glimcher, M.J., Hydroxyl-Groups in Bone-Mineral. Bone, 1995. **16**(5). p. 583–586. DOI: 10.1016/8756-3282(95)00101-I 2.3.2.1

[89] Loong, C.K., Rey, C., Kuhn, L.T., Combes, C., Wu, Y., Chen, S.H., Glimcher, M.J., Evidence of hydroxyl-ion deficiency in bone apatites: An inelastic neutron-scattering study. Bone, 2000. **26**(6). p. 599–602. DOI: 10.1016/S8756-3282(00)00273-8 2.3.2.1

[90] Wachtel, E., Weiner, S., Small-angle x-ray scattering study of dispersed crystals from bone and tendon. J Bone Miner Res, 1994. **9**(10). p. 1651–5. 2.3.2.1

[91] Fratzl, P., Groschner, M., Vogl, G., Plenk, H., Jr., Eschberger, J., Fratzl-Zelman, N., Koller, K., Klaushofer, K., Mineral crystals in calcified tissues: a comparative study by SAXS. J Bone Miner Res, 1992. **7**(3). p. 329–34. 2.3.2.1

[92] Landis, W.J., Glimcher, M.J., Electron diffraction and electron probe microanalysis of the mineral phase of bone tissue prepared by anhydrous techniques. J Ultrastruct Res, 1978. **63**(2). p. 188–223. DOI: 10.1016/S0022-5320(78)80074-4 2.3.2.1

[93] Yu, W.Y., Siu, C.M., Shim, S.S., Hawthorne, H.M., Dunbar, J.S., Mechanical-Properties and Mineral Content of Avascular and Revascularizing Cortical Bone. Journal of Bone and Joint Surgery-American Volume, 1975. **57**(5). p. 692–695. 2.3.2.1

[94] Wall, J.C., Chatterji, S.K., Jeffery, J.W., Influence That Bone-Density and the Orientation and Particle-Size of the Mineral Phase Have on the Mechanical-Properties of Bone. Journal of Bioengineering, 1978. **2**(6). p. 517–526. 2.3.2.1

[95] Sasaki, N., Matsushima, N., Ikawa, T., Yamamura, H., Fukuda, A., Orientation of bone mineral and its role in the anisotropic mechanical properties of bone–transverse anisotropy. J Biomech, 1989. **22**(2). p. 157–64. DOI: 10.1016/0021-9290(89)90038-9 2.3.2.1

[96] Currey, J.D., Effects of Strain Rate, Reconstruction and Mineral Content on Some Mechanical-Properties of Bovine Bone. Journal of Biomechanics, 1975. **8**(1). p. 81-&. DOI: 10.1016/0021-9290(75)90046-9 2.3.2.1

[97] D'Amelio, P., Rossi, P., Isaia, G., Lollino, N., Castoldi, F., Girardo, M., Dettoni, F., Sattin, F., Delise, M., Bignardi, C., Bone mineral density and singh index predict bone mechanical properties of human femur. Connect Tissue Res, 2008. **49**(2). p. 99–104. DOI: 10.1080/03008200801913940 2.3.2.1

[98] Horsman, A.,Currey, J.D., Estimation of mechanical properties of the distal radius from bone mineral content and cortical width. Clin Orthop Relat Res, 1983(176). p. 298–304. 2.3.2.1

[99] Toyras, J., Nieminen, M.T., Kroger, H., Jurvelin, J.S., Bone mineral density, ultrasound velocity, and broadband attenuation predict mechanical properties of trabecular bone differently. Bone, 2002. **31**(4). p. 503–7. DOI: 10.1016/S8756-3282(02)00843-8 2.3.2.1

[100] Nyman, J.S., Reyes, M., Wang, X., Effect of ultrastructural changes on the toughness of bone. Micron, 2005. **36**(7–8). p. 566-82. DOI: 10.1016/j.micron.2005.07.004 2.3.2.4

[101] Robinson, R.A., Bone tissue: composition and function. Johns Hopkins Med J, 1979. **145**(1). p. 10–24. 2.3.2.4

[102] Nomura, S., Hiltner, A., Lando, J.B., Baer, E., Interaction of water with native collagen. Biopolymers, 1977. **16**(2). p. 231–46. DOI: 10.1002/bip.1977.360160202 2.3.2.4

[103] Pineri, M.H., Escoubes, M., Roche, G., Water–collagen interactions: calorimetric and mechanical experiments. Biopolymers, 1978. **17**(12). p. 2799–2815. DOI: 10.1002/bip.1978.360171205 2.3.2.4

[104] Nyman, J.S., Roy, A., Shen, X., Acuna, R.L., Tyler, J.H., Wang, X., The influence of water removal on the strength and toughness of cortical bone. J Biomech, 2006. **39**(5). p. 931–8. DOI: 10.1016/j.jbiomech.2005.01.012 2.3.2.4

[105] Fernandez-Seara, M.A., Wehrli, S.L., Takahashi, M., Wehrli, F.W., Water content measured by proton-deuteron exchange NMR predicts bone mineral density and mechanical properties. J Bone Miner Res, 2004. **19**(2). p. 289–96. DOI: 10.1359/JBMR.0301227 2.3.2.4

[106] Fosse, L., Ronningen, H., Benum, P., Sandven, R.B., Influence of water and fat content on compressive stiffness properties of impacted morsellized bone: an experimental ex vivo study on bone pellets. Acta Orthop, 2006. **77**(1). p. 15–22. DOI: 10.1080/17453670610045641 2.3.2.4

[107] Nyman, J.S., Ni, Q., Nicolella, D.P., Wang, X., Measurements of mobile and bound water by nuclear magnetic resonance correlate with mechanical properties of bone. Bone, 2008. **42**(1). p. 193–9. DOI: 10.1016/j.bone.2007.09.049 2.3.2.4

[108] Ingram, R.T., Clarke, B.L., Fisher, L.W., Fitzpatrick, L.A., Distribution of noncollagenous proteins in the matrix of adult human bone: evidence of anatomic and functional heterogeneity. J Bone Miner Res, 1993. **8**(9). p. 1019–29. 2.3.2.3

[109] Machado do Reis, L., Kessler, C.B., Adams, D.J., Lorenzo, J., Jorgetti, V., Delany, A.M., Accentuated osteoclastic response to parathyroid hormone undermines bone mass acquisition in osteonectin-null mice. Bone, 2008. **43**(2). p. 264–73. DOI: 10.1016/j.bone.2008.03.024 2.3.2.3

[110] Cowles, E.A., DeRome, M.E., Pastizzo, G., Brailey, L.L., Gronowicz, G.A., Mineralization and the expression of matrix proteins during in vivo bone development. Calcif Tissue Int, 1998. **62**(1). p. 74–82. DOI: 10.1007/s002239900397 2.3.2.3

[111] Fisher, L.W.,Termine, J.D., Noncollagenous proteins influencing the local mechanisms of calcification. Clin Orthop Relat Res, 1985(200). p. 362–85. 2.3.2.3

[112] Favus, M.J., Primer of Metabolic bone diseases and disorders of mineral metabolism, Lippincott/Williams & Wilkins: Philadelphia. 1999 2.3.2.3, 2.1

[113] Cowin, S.C., Bone Mechanics Handbook, Informa Healthcare USA, Inc, 2008 2.3.2.3, 2.1

[114] Roy, M.E., Nishimoto, S.K., Rho, J.Y., Bhattacharya, S.K., Lin, J.S., Pharr, G.M., Correlations between osteocalcin content, degree of mineralization, and mechanical properties of C-Carpio rib bone. Journal of Biomedical Materials Research, 2001. **54**(4). p. 547–553. DOI: 10.1002/1097-4636(20010315)54:4%3C547::AID-JBM110%3E3.0.CO;2-2 2.3.2.3

[115] Kavukcuoglu, N.B., Patterson-Buckendahl, P., Mann, A.B., Effect of osteocalcin deficiency on the nanomechanics and chemistry of mouse bones. Journal of the Mechanical Behavior of Biomedical Materials, 2009. **2**(4). p. 348–354. DOI: 10.1016/j.jmbbm.2008.10.010 2.3.2.3

[116] Kavukcuoglu, N.B., Patterson-Buckendahl, P., Mann, A.B., Effect of osteocalcin deficiency on the nanomechanics and chemistry of mouse bones. J Mech Behav Biomed Mater, 2009. **2**(4). p. 348–54. DOI: 10.1016/j.jmbbm.2008.10.010 2.3.2.3

[117] Looker, A.C., Melton, L.J., Harris, T., Borrud, L., Shepherd, J., McGowan, J., Age, gender, and race/ethnic differences in total body and subregional bone density. Osteoporosis International, 2009. **20**(7). p. 1141–1149. DOI: 10.1007/s00198-008-0809-6 2.3.3

[118] Lei, S.F., Deng, F.Y., Li, M.X., Dvornyk, V., Deng, H.W., Bone mineral density in elderly Chinese: effects of age, sex, weight, height, and body mass index. Journal of Bone and Mineral Metabolism, 2004. **22**(1). p. 71–78. DOI: 10.1007/s00774-003-0452-4 2.3.3

[119] Ballard, J.E., Wallace, L.S., Holiday, D.B., Herron, C., Harrington, L.L., Mobbs, K.C., Cussen, P., Evaluation of differences in bone-mineral density in 51 men age 65–93 years: A cross-sectional study. Journal of Aging and Physical Activity, 2003. **11**(4). p. 470–486. 2.3.3

[120] Yao, W.J., Wu, C.H., Wang, S.T., Chang, C.J., Chiu, N.T., Yu, C.Y., Differential changes in regional bone mineral density in healthy Chinese: Age-related and sex-dependent. Calcified Tissue International, 2001. **68**(6). p. 330–336. DOI: 10.1007/s002230001210 2.3.3

[121] Goldman, H.M., Bromage, T.G., Thomas, C.D., Clement, J.G., Preferred collagen fiber orientation in the human mid-shaft femur. Anat Rec A Discov Mol Cell Evol Biol, 2003. **272**(1). p. 434–45. DOI: 10.1002/ar.a.10055 2.3.3

[122] Vincentelli, R., Relation between Collagen Fiber Orientation and Age of Osteon Formation in Human Tibial Compact Bone. Acta Anatomica, 1978. **100**(1). p. 120–128. DOI: 10.1159/000144890 2.3.3

[123] Wang, X., Li, X., Shen, X., Agrawal, C.M., Age-related changes of noncalcified collagen in human cortical bone. Ann Biomed Eng, 2003. **31**(11). p. 1365–71. DOI: 10.1114/1.1623488 2.3.3

[124] Eyre, D.R., Dickson, I.R., Van Ness, K., Collagen cross-linking in human bone and articular cartilage. Age-related changes in the content of mature hydroxypyridinium residues. Biochem J, 1988. **252**(2). p. 495–500. 2.3.3

[125] Sell, D.R., Monnier, V.M., Structure elucidation of a senescence cross-link from human extracellular matrix. Implication of pentoses in the aging process. J Biol Chem, 1989. **264**(36). p. 21597–602. 2.3.3

[126] Dyer, D.G., Blackledge, J.A., Thorpe, S.R., Baynes, J.W., Formation of pentosidine during nonenzymatic browning of proteins by glucose. Identification of glucose and other carbohydrates as possible precursors of pentosidine in vivo. J Biol Chem, 1991. **266**(18). p. 11654–60. 2.3.3

[127] Verzijl, N., DeGroot, J., Thorpe, S.R., Bank, R.A., Shaw, J.N., Lyons, T.J., Bijlsma, J.W., Lafeber, F.P., Baynes, J.W., TeKoppele, J.M., Effect of collagen turnover on the accumulation of advanced glycation end products. J Biol Chem, 2000. **275**(50). p. 39027–31. DOI: 10.1074/jbc.M006700200 2.3.3

[128] Odetti, P., Rossi, S., Monacelli, F., Poggi, A., Cirnigliaro, M., Federici, M., Federici, A., Advanced glycation end products and bone loss during aging. Ann N Y Acad Sci, 2005. **1043**. p. 710–7. DOI: 10.1196/annals.1333.082 2.3.3

[129] Hernandez, C.J., Tang, S.Y., Baumbach, B.M., Hwu, P.B., Sakkee, A.N., van der Ham, F., DeGroot, J., Bank, R.A., Keaveny, T.M., Trabecular microfracture and the influence of pyridinium and non-enzymatic glycation-mediated collagen cross-links. Bone, 2005. **37**(6). p. 825–32. DOI: 10.1016/j.bone.2005.07.019 2.3.3

[130] Aerssens, J., Boonen, S., Lowet, G., Dequeker, J., Interspecies differences in bone composition, density, and quality: potential implications for in vivo bone research. Endocrinology, 1998. **139**(2). p. 663–70. DOI: 10.1210/en.139.2.663 2.4, 2.2, 2.4.2

[131] Beddoe, A.H., Quantitative Study of Structure of Trabecular Bone in Man, Rhesus-Monkey, Beagle and Miniature Pig. Calcified Tissue Research, 1978. **25**(3). p. 273–281. DOI: 10.1007/BF02010781 2.4.1

[132] MacLatchy, L., Muller, R., A comparison of the femoral head and neck trabecular architecture of Galago and Perodicticus using micro-computed tomography (mu CT). Journal of Human Evolution, 2002. **43**(1). p. 89–105. DOI: 10.1006/jhev.2002.0559 2.4.1

[133] Fajardo, R.J., Muller, R., Three-dimensional analysis of nonhuman primate trabecular architecture using micro-computed tomography. American Journal of Physical Anthropology, 2001. **115**(4). p. 327–336. DOI: 10.1002/ajpa.1089 2.4.1

[134] Martiniakova, M., Grosskopf, B., Omelka, R., Vondrakova, M., Bauerova, M., Differences among species in compact bone tissue microstructure of mammalian skeleton: Use of a discriminant function analysis for species identification. 2.4.1

CHAPTER 3

Current Mechanical Test Methodologies

CHAPTER SUMMARY

Chapter 3 intends to explain some fundamentals of the mechanics of materials and gives an overview of current methodologies of mechanical testing of bone in tension/compression, torsion, bending, fracture toughness, fatigue, viscoelasticity, and some recent test schemes. In order to provide sufficient information for tissue engineers to adequately use these test methodologies, general mechanical behavior of materials, formulae for determining the mechanical properties from the experimental data, and specific issues in bone testing are discussed for these tests.

3.1 INTRODUCTION

Mechanical testing has been a direct way to evaluate the mechanical behaviors of bone tissues [2–9] and bone substitute materials [5,10–14]. The methods of mechanical testing are based on fundamental principles of the mechanics of materials. Forces/stress and displacement/strain are key variables measured through mechanical testing. When bone tissue is subjected to external forces, the response of the tissue depends on a number of factors, such as the type of loading (tension, compression, shear, bending, or combinations), the nature of loading (monotonic vs. cyclic loading), the specimen size (macro, micro or nano length scale), and the loading rate (impact vs. static).

There are many different types of standard and customized mechanical tests, each of which provides unique information pertaining to a specific mechanical behavior of materials under the prescribed test condition. This chapter intends to provide a brief summary of the procedures for specimen preparation, test setup, stress–strain measurements, and interpretation of experimental results for the mechanical tests that are most often utilized in bone biomechanics. It starts with the description of mechanical tests for macroscopic specimens, followed by the micro and nanoscopic mechanical tests. The objective of this chapter is to help understand basics of mechanical testing methodologies and how to use them adequately in their research.

3.2 DEFINITION OF MECHANICAL PROPERTIES

3.2.1   BASIC CONCEPTS OF MECHANICS OF MATERIALS

Mechanics of materials is a branch of engineering science that studies the physical behavior of a deformable body under load and the relationship between the deformation of the body and the external load applied. There are three basic types of forces: tension, compression, and shear. Tension and compression forces tend to lengthen and shorten the deformable body (or structure), respectively, whereas shearing forces tend to cause sliding of one part over the other in the body. The ratio of the applied force to the corresponding deformation is usually defined as the stiffness of the body. The maximum force that the body or structure can sustain is referred to as the failure load. Moreover, the maximum work done by the force to the deformed body is defined as the work to failure. Such measurements are often employed in the mechanical tests of whole bones. However, it should be understood that these parameters describe a combined effect of the geometry of the body and the type of material out of which it is made. For example, if a wood bar is thick enough, it could be stronger and stiffer than a steel wire although intuitively we know that as a material steel is much stronger and stiffer than wood. In order to measure the intrinsic mechanical properties of a material, it becomes necessary to remove the geometrical contribution from the measurements. This can be done by dividing the force applied to the body by the area over which it is applied (i.e., stress) or dividing the amount of deformation by the original dimension of the body (i.e., strain). In this context, the intrinsic material properties of a body can be measured by finding the relationship between the stress (i.e., the force per unit area) and strain (i.e., the percentage deformation per unit length) in the body. In the simplest definition, strain is the normalized deformation by the original length of the subject, and the stress is the normalized load by the area in the projection plane that is perpendicular to the load direction.

The applied load(s) and the corresponding displacement(s) are the common measurable parameters in mechanical tests of materials. Based on the load and displacement readings, a formula derived from the mechanics of materials to calculate the stresses and strains. By plotting the stress with respect to the strain, a strain-stress curve can be obtained (Figure 3.1) from which the mechanical properties of the material can are measured.

In general, the behavior of a material in quasi-static mechanical tests may experience the following subsequent stages: elasticity, post-yield deformation, and failure. Other behaviors, such as fatigue and viscoelasticity, are usually observed in dynamic loading modes.

3.2.2   ELASTICITY

Elastic behavior characterizes reversible deformation, i.e., the original dimensions of the specimen can be recovered after removing the applied load. It can be described mathematically by the generalized Hooke's Law,

$$\sigma_i = C_{ij}\varepsilon_j \tag{3.1}$$

where the six stress components (σ_{1-6}) are related to the six strain components (ε_{1-6}) through a stiffness matrix consisting of 36 elastic constants (C_{ij}) which reduce to 21 independent constants by the symmetry of properties. Orthotropic materials have properties that differ in each of three

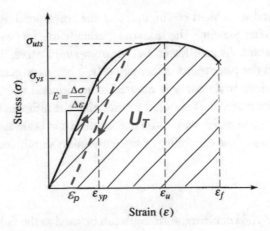

Figure 3.1: Simplified stress-strain diagram.

perpendicular directions, thus reducing the independent elastic constants to nine (9) in this case. Transversely isotropic materials have properties that are the same in every direction about an axis of symmetry and are described by five (5) independent elastic constants. Isotropic materials possess the same properties in all directions and their elastic properties can be described by only two (2) elastic constants: i.e., the elastic modulus (E) and the shear modulus (G). In this case, Equation (3.1) reduces to two equations:

$$\sigma = E\varepsilon \tag{3.2}$$
$$\tau = G\gamma . \tag{3.3}$$

The modulus of elasticity, E, is the ratio of stress to strain when deformation is linearly elastic, i.e., the slope of the initial linear region of the stress-strain curve (Figure 3.1). The shear modulus, G, is the slope of the linear elastic region of the shear stress (τ) and strain (γ) curve.

A third parameter, Poisson's ratio (ν), relates the lateral to axial strains with respect to the loading direction through the following equation

$$\nu = -\frac{\varepsilon_x}{\varepsilon_y} = -\frac{\varepsilon_y}{\varepsilon_z} . \tag{3.4}$$

In the isotropic case, Poisson's ratio is not an independent parameter as it is related to the elastic and shear moduli through the following equation

$$E = 2G(1+\nu) . \tag{3.5}$$

3.2.3 POST-YIELD PROPERTIES

Yielding defines the transition of a material's behavior from elastic to plastic state. The yield point on the stress-strain curve is usually determined using the so-called 0.2% strain offset method. The strain at yielding is referred to as yield strain, ε_{ys}, and the corresponding stress is defined as yield stress, σ_{ys} (Figure 3.1). After yielding, the material gradually loses its capacity to sustain the load as the deformation progresses. As manifested in the stress-strain curve, the stress starts to level off with increasing strain and the permanent deformation (ε_p) begins to accumulate until failure. The maximum stress on the stress-strain curve is referred to as ultimate stress (σ_{ut}) and is often called the ultimate strength of the material. It should be noted that the ultimate strength is not necessarily the stress at failure of the material. However, the maximum strain is always the strain at failure (ε_f). σ_{ut} and ε_f are the measures of the capability of the material to sustain load and deformation up to the state of failure.

3.2.4   FAILURE

As aforementioned, either yield or ultimate strength can be used as the failure criterion of materials. This gives a measure of the maximum stress that can be applied to the material without causing failure. The yield strength can be used when yielding is not allowed for the material. Otherwise, the ultimate strength is usually used as the failure criterion for the material.

As a measure of failure property, toughness is a parameter denoting the total energy dissipation/absorption per unit volume up to failure of a material. It can be readily determined by the area under the stress-strain curve of the material (i.e., U_T shown in Figure 3.1). Since toughness reflects the capacity of a material to absorb energy until failure, it is considered as one of the best parameters to estimate the fragility of bone [15].

In addition to toughness, there is another frequently measured property that quantifies resistance of materials to failure: i.e., fracture toughness. Specifically, fracture toughness measures the resistance of a material to propagation of an existing crack. Linear elastic fracture mechanics is usually applied to determine the fracture toughness of materials using two parameters: the critical stress intensity factor (K_c MNm$^{-3/2}$) and the critical strain energy release rate (G_c J/m^2). As the stress field at the vicinity of the crack tip can be defined by a stress intensity factor, the critical stress intensity factor (K_c) is simply the critical value of the stress intensity for which crack extension occurs. In general, K_c can be calculated using the following equation:

$$K_c = Y \cdot \sigma_c \sqrt{\pi a} \qquad (3.6)$$

where, Y is a function of specimen and crack geometry, a is the crack length, and σ_c is the critical applied stress to the specimen at which the crack starts to propagate.

The strain energy release rate (G_c) is the surface energy dissipated during fracture per unit area of the newly created crack faces. Based on linear elastic fracture mechanics, it is related to K_c and expressed as

$$G_c = \frac{K_c^2}{E}. \qquad (3.7)$$

A crack can be loaded in three fundamental ways known as modes (Figure 3.2), each of which represents the relative crack surface displacement in a different plane, which are known as

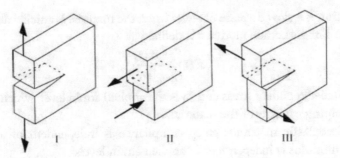

Figure 3.2: The three modes of load to a crack: opening mode is known as Mode I, whereas sliding and tearing modes are represented by Modes II and III.

opening (Mode I), sliding (Mode II), and tearing (Mode III) modes, respectively. In addition, fracture resistance curves (R-curves) can be obtained by plotting G_c with respect to the crack length of a pre-cracked specimen to predict the behavior of crack propagation.

3.2.5 FATIGUE

Fatigue is a form of failure in which a material subjected to dynamic or fluctuating stresses fails at much lower stress or strain levels than the failure stress/strain in the static loading. The fatigue properties of a material are experimentally determined by subjecting a specimen to a cyclic loading scheme at a certain stress amplitude and counting the number of cycles to failure. The resulting plot of the stress amplitude "S" versus the logarithm of the number of cycles to failure (N_f) is known as an S-N curve. N_f is referred to as fatigue life at a specified stress level, representing the number of loading cycles at which the material fails.

Additionally, degradation of mechanical properties of bone induced by fatigue can be assessed by monotonically testing specimens that have been subjected to previous bouts of fatigue loading and comparing the results to the properties of non-fatigued specimens.

3.2.6 VISCOELASTICITY

Viscoelasticity is a phenomenon of time-dependent stress-strain relationship for some materials under a dynamic load. Such a relationship is also frequency dependent. Typical viscoelastic behavior of a material includes creep, stress relaxation, and mechanical damping.

Creep is a slow change of deformation when a material is quickly loaded to a constant force. The creep compliance $J(t)$ is a parameter used to determine the damping capacity of the material (see Equation (3.16)) and is defined as time-dependent strain $\varepsilon(t)$ of a material divided by the applied stress level σ_0.

$$J(t) = \frac{\varepsilon(t)}{\sigma_0} . \tag{3.8}$$

Stress relaxation is a slow decrease of stress when the material is quickly deformed to and held at a constant strain. The relaxation modulus is defined as

$$E(t) = \frac{\sigma(t)}{\varepsilon_0} \tag{3.9}$$

where $\sigma(t)$ is the time-dependent stress and ε_0 is the applied strain level, which can also be used to determined the damping capacity of the material.

For linear viscoelastic materials, creep compliance is independent of applied stress levels whereas relaxation modulus is independent of applied strain levels.

If cyclic loading is applied to a viscoelastic material, a phase lag between stress and strain (Figure 3.3) represents a dissipation of mechanical energy. The stress can be expressed as

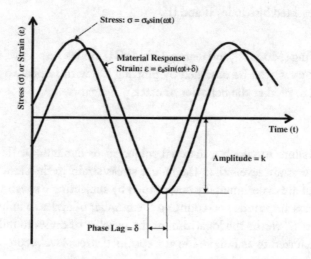

Figure 3.3: Phase lag of a viscoelastic material between a sinosoidal stress and a sinosoidal strain.

$$\sigma = \sigma_0 \sin \omega t \tag{3.10}$$

where σ, σ_0, ω, and t are the applied stress, the maximum stress, the frequency of oscillation, and time, respectively. The strain of the visoelastic material is expressed as

$$\varepsilon = \varepsilon_0 \sin(\omega t + \delta) \tag{3.11}$$

where ε is the strain at time t, ε_0 is the maximum strain, δ is the phase angle difference between the applied stress and the resultant strain.

For viscoelastic materials, the modulus can be expressed in a complex form. The complex modulus (E^*) has two terms, one is related to the storage of energy (i.e., storage modulus, E') and the other to the dissipation of energy (i.e., loss modulus, E'').

$$E^* = E' + i E'' . \tag{3.12}$$

The storage modulus is define as

$$E' = \frac{\sigma_0}{\varepsilon_0} \cos \delta \qquad (3.13)$$

and the loss modulus is defined as

$$E'' = \frac{\sigma_0}{\varepsilon_0} \sin \delta . \qquad (3.14)$$

In addition, the loss tangent, a parameter representing the dampling capacity of the material, is calculated as the ratio of the loss modulus to storage modulus

$$\tan \delta = \frac{E''}{E'} . \qquad (3.15)$$

Viscoelastic behavior of a material can be described by storage modulus, loss modulus, and loss tangent. Only two of these parameters are independent. The complex modulus in the frequency domain is related to the creep compliance and stress relaxation modulus in the time domain through Fourier transformation.

The loss tangent of a linearly viscoelastic material is related to the slope of the creep curve by the following approximation:

$$\tan \delta \approx -\frac{\pi}{2} \frac{d \ln J(t)}{d \ln t}\bigg|_{t=1/\omega} . \qquad (3.16)$$

The loss tangent is also related to the slope of the stress relaxation curve by the following approximation:

$$\tan \delta \approx -\frac{\pi}{2} \frac{d \ln E(t)}{d \ln t}\bigg|_{t=1/\omega} . \qquad (3.17)$$

3.3   MECHANICAL TESTING OF BULK BONE TISSUES

Since bone is an anisotropic material, its mechanical properties vary with directions for both cortical [16–19] and trabecular bones [18], [20–23]. Therefore, the type of loading has great influence on the mechanical behavior of bone at the macroscopic or bulk tissue level. In addition, the mechanical behavior of bone is also loading-rate or strain-rate dependent [24–31]. Most mechanical testing of bulk bone tissue is similar to mechanical testing of other materials. In this section, both monotonic loading tests (tension, compression, torsion, bending, and fracture toughness test) and dynamic loading tests (progressive loading, fatigue loading, and dynamic mechanical analysis) are discussed.

3.3.1   TENSILE/COMPRESSIVE TEST

Both tension and compression tests are standard uniaxial tests in which test specimens are loaded in a mechanical testing machine under an axial load in either load or displacement/strain control.

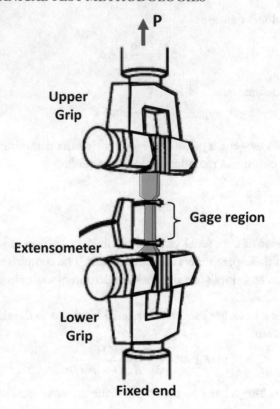

Figure 3.4: Typical setup for a tensile test of bone.

3.3.1.1 Tensile Specimen Preparation and Setup

Figure 3.4 shows a typical setup for a tensile test. Tensile test specimens are usually machined into dog-bone shaped strips or cylinders. These specimens have two gripping regions at both ends that taper into a uniform gage region with a reduced cross-sectional area. The gripping regions are clamped tightly in the upper and lower grips to ensure that no sliding occurs during loading between the specimen and the clamps. In such a setup, a uniform strain and a much higher stress can be achieved in the gage region of the specimen. An extensometer is usually attached in the gage region to measure the elongation of the specimen over the gage length (L_o) preset by the extensometer. If an additional extensometer is attached to the gage region to measure the change in the width of the gage, the Poisson's ratio in tension can be estimated by combining the measurement with the length change in the gage region.

It is noteworthy that it is often hard to grip spongy-like trabecular bone specimens in tensile tests because the specimens may be easily crushed under the large gripping forces. In this case, the two ends of the specimen are usually embedded in plastic resins to ensure that the specimen be securely clamped in the grips without failure. In addition, in many cases it is hard to measure the deformation using regular extensometers for trabecular bone specimens. As an alternative, commercially available optical extensometers can serve for the purpose.

3.3.1.2 Compressive Test Specimen Preparation and Setup

Compressive test specimens are usually machined into short cylindrical or square columns (Figure 3.5). In order to avoid buckling during the compression test, the aspect ratio (height over

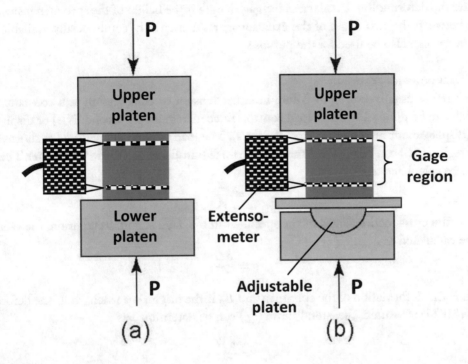

Figure 3.5: Typical compression specimen and loading configurations: (a) two fixed platens, (b) one fixed platen and one adjustable platen.

width/diameter) of the specimen is suggested to be lower than 2.0 [24–31] [32]. Figure 3.5(a) shows the schematic representation of a standard setup for compression tests. Before inserting the bone specimen in between the upper and lower platens, the specimen needs to be polished to ensure the parallelism of its two ends. By doing so, the uniform engagement between the surfaces of the platens and specimen ends can be achieved. For further improvement, a self-adjustable platen with a spherical joint can be introduced to compensate the misalignment of the specimen for full engagement between the platens and the specimen ends as shown in Figure 3.5(b). Such self-adjustable platens are commercially available. An extensometer is usually attached to the specimen to measure the shortening of the specimen over the gage length (L_o) preset by the extensometer. If an additional extensometer is attached to the specimen to measure its lateral deformation, the Poisson's ratio in compression can be estimated combining the measurement of the shortening in the gage region.

As aforementioned, for testing trabecular bones the two ends of the specimen should be embedded in plastic resins to ensure that the specimen be securely engaged between the two platens and that a uniform deformation is applied to the specimen. In the compression test of trabecular bone specimens, it may not be suitable for directly attaching the extensometer to the specimen surface, which is irregular and porous. To avoid the difficulty, the extensometer can be attached to the platens to measure the deformation. In this case, the gage length is the height of the specimen instead of the distance between the two edges of the extensometer. Alternatively, commercially available optical extensometers can also be used for the purpose.

3.3.1.3   Experimental Procedure

The specimen is usually loaded to failure in either tension or compression at a constant loading rate, which can be defined as either load control (force increment per second: N/s) or displacement control (displacement increment per second: m/s). The load (P) and elongation (or shortening) in the gage region (δL) can be recorded through the force transducer and the extensometer. The tensile stress can be calculated as

$$\sigma = \frac{P}{A} \tag{3.18}$$

where A is the cross sectional area of the specimen in the gage region. In addition, the axial strain (ε) can be calculated as

$$\varepsilon = \frac{\delta L}{L_o} \tag{3.19}$$

where δL is the deformation of the specimen and L_o is the original gage length. If the deformation in width (δW) is measured, the lateral strain (ε_L) can be determined as

$$\varepsilon_L = \frac{\delta W}{W_o} . \tag{3.20}$$

Then, Poisson's ratio in tension can be calculated as

$$\nu = -\frac{\varepsilon_L}{\varepsilon} . \tag{3.21}$$

3.3.1.4   Determination of Mechanical Properties

First, the stress-strain curve can be obtained by plotting the stress with respect to the strain as shown in Figure 3.1. From the stress-strain curve, the following mechanical properties can be obtained for the test specimen: elastic modulus, yield stress/strain, ultimate stress, strain at failure, and toughness.

The elastic modulus is determined as the slope of the initial linear elastic part of the stress-train curve

$$E = \frac{\Delta \sigma}{\Delta \varepsilon} . \tag{3.22}$$

The yield stress is determined using the 0.2% strain offset method. Briefly, a straight line is drawn starting from 0.2% strain on the strain axis and parallel to the initial linear region of the stress-strain curve. The intercept of this line with the stress-strain curve is defined as the yield point. The corresponding stress and strain at the yield point are used as yield stress, σ_{ys} and yield strain, ε_{ys}, respectively, (Figure 3.6).

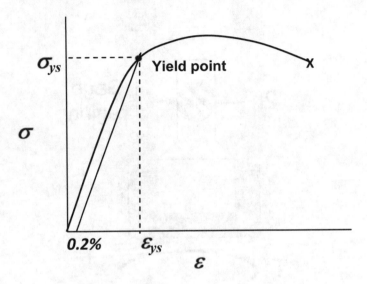

Figure 3.6: Schematic representation for determining the yield point using 0.2% strain offset method.

The ultimate strength (σ_{ut}), toughness (U_T), and strain at failure (ε_f) are determined from the stress-strain curve as described earlier as shown in Figure 3.1.

3.3.2   TORSION (SHEAR) TEST

Torsion test is a mechanical testing method to determine the shear properties of materials. Torsional specimens usually have solid circular or annular cross section with a reduced gage region. It is clamped in grips at each end and subjected to torsion in a testing machine (Figure 3.7). The specimen is loaded to failure at a constant loading rate, which can be in either torsional load control or angular displacement control.

Based on the mechanics of materials, the maximum shear stress (τ) and strain (γ) can be calculated using the following equations,

$$\tau = \frac{T r}{I_p} \tag{3.23}$$

$$\gamma = \frac{\phi r}{L} \tag{3.24}$$

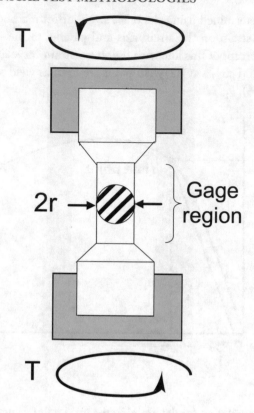

Figure 3.7: Typical torsion specimen and loading configuration.

where, T is the applied torque, r is the radius of the gage region of the specimen, ϕ is the twist angle, I_p is the polar moment of inertia of the cross-section, and L is the gage length. The torque can be measured using a torque transducer and the twist angle in the gage region can be measured using a twist angle sensor. The shear stress-strain curve can be obtained. From the curve, the shear modulus (G) is given by

$$G = \frac{\Delta \tau}{\Delta \gamma}. \tag{3.25}$$

Similarly, shear yield strength (τ_{ys}), ultimate shear strength (τ_{ut}), and shear strain at failure (γ_f) can be estimated following the procedure aforementioned in tension/compression tests.

3.3.3 3-POINT AND 4-POINT BENDING TEST

There are two common types of bending configurations: three-point bending (Figure 3.8(a)) and four-point bending (Figure 3.8(b)). Bending specimens are usually in the shape of prismatic beams. Since specimen preparation and experiment setup are relatively simple compared with other tests,

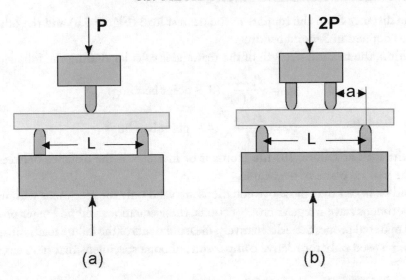

Figure 3.8: Typical bending specimen and loading configurations in (a) three-point bending and (b) four-point bending configuration.

bending tests have been widely used for testing of biological tissues and materials. However, since a bending specimen experiences tension, compression, and shear all at once during testing, the flexure modulus and the flexure strength are usually not equivalent to the elastic modulus and the strength defined earlier. Such differences can be clearly observed when bone specimens are tested.

In order to reduce the effect of shear on the test, the length of the beam should be much longer than the cross-sectional dimensions of the specimen. In 3-point bending tests, the test specimen is placed on two supports and a force is applied at the center of the beam. In this case, the maximum bending moment is applied at the center of the specimen. In 4-point bending test, the specimen is placed on two supports and a couple of two equal forces are symmetrical applied on the top of the specimen. In this case, the maximum bending moment is uniformly applied to the center portion of the specimen between the two forces (Figure 3.8(b)).

Based on beam theory, the bending (flexure) modulus is determined using the load-displacement curve. For 3-point bending, it is expressed as

$$E = \frac{48L^3}{I} \frac{\Delta P}{\Delta \delta} \qquad (3.26)$$

where, $\Delta P / \Delta \delta$ is the slope of the initial linear region of the load-displacement curve, I is the moment of inertia of the cross-section, B is the thickness, W is the width, L is the length of support span of the specimen. For 4-point bending, it is expressed as

$$E = \frac{a^2}{6I} (3l - 4a) \frac{\Delta P}{\Delta \delta} \qquad (3.27)$$

where, a is the distance from the support to the nearest load (Figure 3.8) and the other parameters are the same as defined in 3-point bending.

In addition, the flexural strength of the material is calculated using the following equations,

$$\sigma_c = \frac{P_c L c}{4I} \quad (3 - \text{point bending}) \tag{3.28}$$

$$\sigma_c = \frac{P_c a\, c}{I} \quad (4 - \text{point bending}) \tag{3.29}$$

where, P_c is the load at failure, I is the moment of inertia, c is the distance between the bottom surface and the neutral plane of the beam.

It should be noted that the above equations are valid only for specimens with uniform cross-sections. If specimens have irregular cross-sections, these equations will no longer provide accurate but estimated material properties. Alternatively, measurements of the failure load, stiffness, and work to fracture can be used only for relative comparisons of bone specimens that have similar shapes.

3.3.4   PROGRESSIVE LOADING

In addition to monotonic tests, mechanical tests utilizing diagnostic cycles [33–35] and progressive loading schemes [15,36,37] have found increasing use in characterizing the post-yield behavior of bone. These protocols involve loading a bone specimen in cycles with an increasing displacement until failure of the specimen. Figure 3.9 illustrates the progressive loading scheme [15,36,37] in which

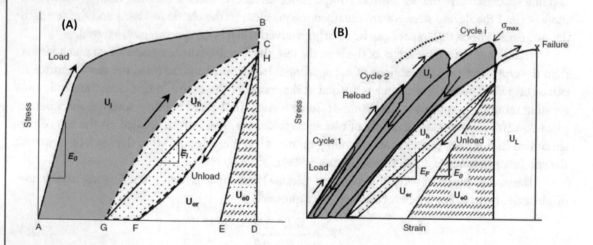

Figure 3.9: (A) An example of loading-unloading-reloading scheme with rest insertion (dwell) used to partition the energy dissipation of bone. (B) Progressive yield scheme in which the specimen is load to failure in cycles of increasing strain and the energies quantified at the last cycle (B) [15].

the energy absorbed during the post-yield deformation is partitioned into three components: the released elastic strain energy (U_{er}), the hysteresis energy (U_h), and the permanent energy (U_p) at each cycle. Each of these components reflects different processes occurring within the bone. U_{er} has been shown to be associated with modulus degradation induced by accumulation of microdamage [38], U_h with the elevated viscoelastic behavior, and U_p with the plastic deformation of the tissue [38].

In addition, strain-dependent changes in the elastic modulus, plastic strain, and viscous response of bone can be quantified using this progressive loading scheme. These measurements can provide detailed information regarding the evolution of modulus degradation, plastic deformation, and energy dissipation during the post-yield deformation of bone. It is impossible to acquire such information through traditional tests. The progressive loading scheme can be applied to tension, compression, and torsion/shear tests.

3.3.5   FRACTURE TOUGHNESS TEST

The most common fracture toughness tests on bone specimens are compact tension and single-end-notched bending tests. Due to the limitation of bone stocks, fracture toughness test specimens are usually prepared from human and large animals.

3.3.5.1   Compact Tension Test

Figure 3.10 shows a typical compact tension specimen for opening loading mode (Mode I) fracture toughness tests. A sharp notch is introduced on one side of the specimen along the center plane and

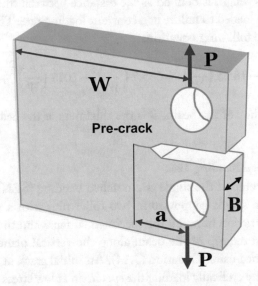

Figure 3.10: A typical compact tension specimen for fracture toughness testing.

two pinholes are drilled on both sides of the notch. In general, the ratio of height to width of the specimen is about 0.8~0.9. The thickness of the specimen (B) needs to be determined using the

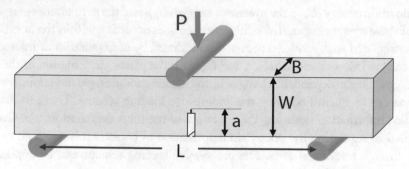

Figure 3.11: Schematic representation of a single end notched bending test.

following equation,

$$B = 2.5 \cdot \left(\frac{K_c}{\sigma_{ys}}\right)^2 \tag{3.30}$$

where, K_c is the fracture toughness and σ_{ys} is the yield strength of the material, respectively. A pair of splitting forces is applied to the crack through the pins inserted in the two pinholes. Prior to testing, an initial crack at the notch tip is first introduced by cyclically loading the specimen at low stress amplitude. The crack length is defined as the distance between the crack tip and the loading line. Then, the specimen is loaded to failure at a constant loading rate. The fracture toughness (K_c) is then calculated using the following equation:

$$K_{Ic} = \frac{P_c}{B\sqrt{W}}\left[29.6\left(\frac{a}{W}\right)^{\frac{1}{2}} - 185.5\left(\frac{a}{W}\right)^{\frac{3}{3}} + 655.7\left(\frac{a}{W}\right)^{\frac{5}{2}} - 1017\left(\frac{a}{W}\right)^{\frac{7}{2}} + 638.9\left(\frac{a}{W}\right)^{\frac{9}{2}}\right] \tag{3.31}$$

where, P_c is the load at failure of the beam, B is the thickness of the beam, and W is the width of the beam.

3.3.5.2 Single-End Notched Bending Test

Figure 3.11 illustrates the setup of the single end notched bending (SENB) test. A prismatic beam with a notch on its bottom side is supported in two roller pins with a support span of L. These pins may help remove the friction between the specimen surfaces and the fixture. Then, a load (P) is applied through a pin on the top of the beam along the vertical plane. Similarly, the thickness of the specimen is determined using Equation (3.14). An initial crack at the notch tip needs to be introduced prior to testing by cyclically loading the specimen at low stress amplitude. The specimen is loaded to failure at a constant loading rate. The fracture toughness of the material is determined using the following equation,

$$K_{Ic} = \frac{P_c L}{B\,(W)^{3/2}}\left[2.9\left(\frac{a}{W}\right)^{\frac{1}{2}} - 4.6\left(\frac{a}{W}\right)^{\frac{3}{3}} + 21.8\left(\frac{a}{W}\right)^{\frac{5}{2}} - 37.6\left(\frac{a}{W}\right)^{\frac{7}{2}} + 37.7\left(\frac{a}{W}\right)^{\frac{9}{2}}\right] \tag{3.32}$$

where, P_c is the load at failure of the beam, a is the crack length, L is the length of support span, B is the thickness, and W is the width of the beam.

3.3.6   CHARPY AND IZOD IMPACT TESTS

In addition to the fracture toughness tests, Charpy and Izod impact tests are often utilized to measure the impact failure energy of a material (Figure 3.12). This measurement is referred to as the notch

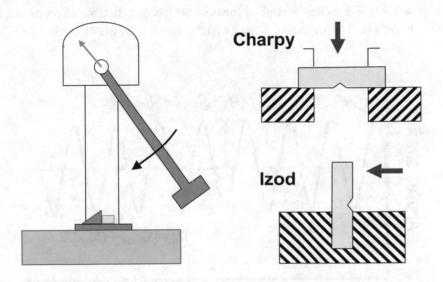

Figure 3.12: Schematic representation of Charpy and Izod tests.

toughness. In both tests, a notched rectangle bar of the material is struck by a weighted pendulum released from a known height, h_1. A knife-edge on the pendulum strikes and fractures the specimen at the notch. After striking the specimen, the pendulum continues to travel upward to a height, h_o, which is lower than the release height. The energy dissipated in fracturing the specimen is computed from the differences in height and is a measure of the impact energy (E_{impact}).

$$E_{impact} = m \cdot g \cdot (h_1 - h_o) \ . \tag{3.33}$$

Where, m is the mass of the pendulum, g is the gravitational acceleration, h_1 is the height of the center of mass of the pendulum prior to the impact, and h_o is the height of the center of mass of the pendulum after the impact. Unlike K_c, G_c, and toughness, impact tests are qualitative in nature and are mainly used in making relative comparisons between samples.

3.3.7   FATIGUE TEST

Tensile, compressive, bending and torsional loading can be used to measure fatigue properties of a material through repetitive loading. Preparation of specimens for fatigue testing can be referred to previous sections depending on the loading mode that is selected for the test.

Fatigue testing requires a test machine capable of applying cyclic loading (e.g., a servo-hydraulic testing machine or an electromagnetic testing instrument). In fatigue tests, the mean stress (σ_m) and applied stress amplitude (σ_a) have the most significant effect on the fatigue life of a material (Figure 3.13). For example, applied mean stress has considerable effect on the fatigue life of cortical bone. At a low mean stress ($\sigma_m = 24$ MPa), the fatigue life of cortical bone is about 37

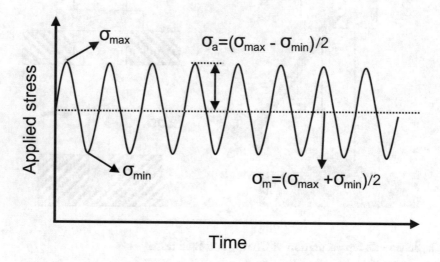

Figure 3.13: A typical fatigue loading curve under stress control. σ_{max} is the maximum stress and σ_{min} is the minimum stress during loading. σ_m is the mean stress. σ_a is the amplitude of applied stress.

million cycles [39]. However, If the applied mean stress is increased to 60 MPa, cortical bone would quickly fail at about 21,000 cycles. Therefore, a fatigue test needs to be designed carefully to achieve objectives [40].

During fatigue testing, dehydration may significantly change the fatigue behavior of bone. Therefore, an irrigation system or a water bath is essential in keeping bone samples wet and lubricated. In the irrigation system, normal saline buffered with calcium chloride is important for maintaining stiffness during the fatigue testing [41].

An S-N diagram, depicting the applied stress amplitude as a function of fatigue life (i.e., the number of cycles up to failure, N_f), can be used to represent the results of fatigue tests (Figure 3.14).

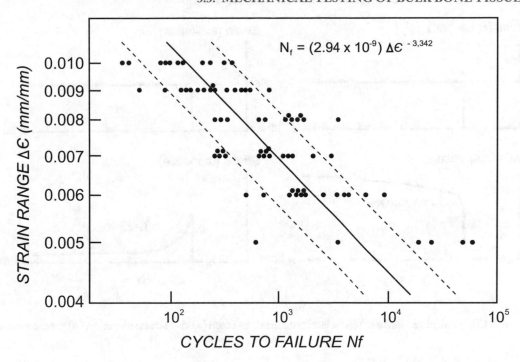

Figure 3.14: An example of S-N curves of cortical bone. Adapted from Carter et al., J Biomech, 1981, 14:461-470.

3.3.8 VISCOELASTIC TEST

Viscoelastic test can be applied to tensile, compressive, bending and torsional loading modes to determine the viscoelastic properties of materials. Most common viscoelastic tests include creep and stress relaxation tests. In addition, dynamical mechanical analysis can also be used to obtain viscoelastic properties such as storage modulus, loss modulus, and loss tangent, but this requires more complex testing devices.

In a creep test, stress is suddenly applied to the specimen at time zero, and the strain is measured as a function of time afterwards. In a stress relaxation test, the strain is held constant and the stress decreases with time. Both creep tests and stress relaxation tests can be conducted on a commercial mechanical testing system (Figure 3.15).

Dynamical mechanical analysis of a material can be conducted by a dynamic mechanical analyzer (DMA). The oscillating force is generated by a force motor and applied to the specimen through a central rod suspended in a magnetic linear bearing (Figure 3.16). The deformation of the specimen is detected by a linear variable differential transformer. An environmental system is used to implement precise temperature control by heating the system with a furnace and by cooling the furnace heat sink with a cooling system through which tap water is constantly running. The DMA

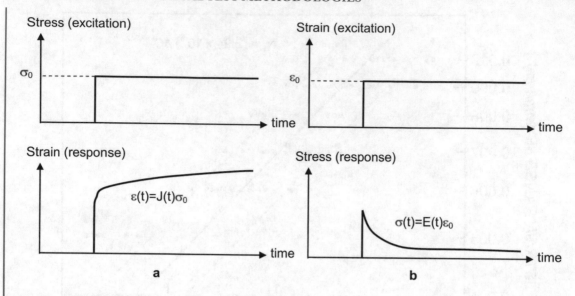

Figure 3.15: Typical response of viscoelastic materials in creep (a) and stress relaxation (b) measurements.

supplies an oscillatory force, which generates a sinusoidal strain (Figure 3.16), and it measures the peak deformation of the specimen. The time lag between peak load and peak deformation of the specimen is calculated to determine the viscoelastic properties of a material, including storage modulus, loss modulus, and loss tangent.

Linear viscoelasticity has been used to determine viscoelastic properties from creep and stress relaxation experiments. Ideal elements are used to describe the creep and stress relaxation process. The creep process is described by a Voigt element, whereas the stress relaxation can be characterized by a Maxwell element. Both elements are called a Debye-type relaxation because an exponential decay function ($e^{t/\tau}$) is included in these elements. For a creep test, the Debye model is expressed as

$$J(t) = J_2 - J_1 e^{-\frac{t}{\tau_c}} \tag{3.34}$$

where τ_c is a retardation time, J_1 and J_2 are constants. The Debye model for stress relaxation as a function of time is

$$E(t) = E_2 + E_1 e^{-t/\tau_r} \tag{3.35}$$

where τ_r is a relaxation time, E_1 and E_2 are constants.

3.4   MICRO-MECHANICAL TESTING METHODOLOGIES

Determining the mechanical properties of micro bone specimens involves challenges in sample preparation and testing. Protocols have been developed that allow for testing bone specimens of

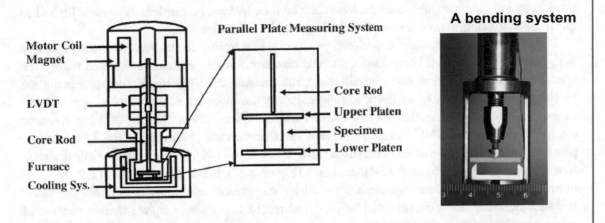

Figure 3.16: Schematic of PerkinElmer DMA-7e showing compression testing and bending testing.

single osteons and interstitial bone tissue from cortical bone. Additionally, it is now possible to perform mechanical tests of single trabeculae. These protocols are often similar in principal to their macroscopic counterparts. However, special procedures are required to identify and produce samples from the tissue of interest. Further, specially designed testing devices and miniaturized fixtures are often required to hold the specimens, apply the loads, and measure the resulting forces and displacements. Tests such as these have greatly expanded our understanding of the mechanical properties of bone at microscopic length scale. Additionally, the information obtained from studies at the micro level gives perspective to the relationship of structures at this scale to higher and lower levels of the bone hierarchy.

Conducting a micro-mechanical test is also challenging. The production of a stress-strain diagram requires knowing the original sample dimensions and acquiring measurements of the force and displacement throughout the test. Due to the small sample size, obtaining dimensions usually entails measuring the sample on a micrometer stage or scaled photograph. Applying the force and obtaining force and displacement measurements requires doing so over a sufficient range and at a sufficient level of resolution to make the results meaningful. The applied forces must often be very low (on the order of grams), which requires sensitive mechanical testing/measuring systems and careful handling during the test. For measuring deformation of such small samples, it is often necessary to use special equipment (e.g., optical devices) to track the movement of the fixtures or points on the sample.

Early work in the area of micro-tensile testing includes manually testing hand-fabricated osteon specimens [42]. In these tests, osteons are identified using polarized light microscopy and dissected from a thin slab of cortical bone using a very sharp needle. The final specimen consists of a rectangular gage region (50μm wide by $20\sim50\mu$m thick by $100\sim300\mu$m long) with grip regions at both ends. The specimen ends are glued to upper and lower jaws of a test fixture and tested in tension by hanging a weight from the lower jaw under a hydrated condition. The test is limited to quasi-static loading, and only ultimate tensile strength can be obtained.

Some of the difficulties in specimen production and testing can be overcome using specially designed devices [43, 44]. For example, a microwave extensometer system can be used to grip the specimen ends, apply quasi-static tensile loads, and measure the small displacements experienced by the gage region [43]. In addition, improvements in the specimen acquisition and testing techniques have been made in the study of osteons and interstitial tissues [44]. By utilizing eccentric coring procedures and CNC techniques, *a priori* identification and production of standardized samples from secondary osteons and interstitial tissue is developed [45]. By utilizing specially designed fixtures and optical displacement measurements (Figure 3.17), it becomes possible to test the specimen using highly accurate, commercially available mechanical testing systems. Using the above techniques, stress-strain curves could be produced to obtain a more complete characterization of micro-mechanical properties.

Micro-mechanical testing protocols have been developed to test trabecular tissue [46], [47,48] and obtain micro-mechanical properties for bone in compression [45], bending [47, 49, 50], shear [45], torsion [51], and fatigue [48]. Using hand dissection procedures and optical strain measurements, the post-yield properties of single trabeculae in tension were determined [46]. Through modification, the microwave extensometer technology can be utilized to determine the compression [52], bending [49], and shear [45] properties of single osteons. Miniaturized fixtures and off-the-shelf components are often necessary to determine the mechanical properties of secondary osteons and interstitial tissues in compression [53] and bending [47,50]. The fatigue test of machined cortical and trabecular micro-specimens is also available [48].

3.5   NANO-INDENTATION METHODOLOGIES

Nanoindentation testing methods have been utilized to determine the material properties of sub-microstructural features of mineralized tissues such as cortical and trabecular bone, enamel, and dentin. Nanoindentation technologies utilize a rigid indenter to penetrate a material's surface, thereby inducing local surface deformation (Figure 3.18). The force and penetration depth are recorded continuously during loading, holding, and unloading portions of a test. Curves of force versus displacement are generated from which the material properties are determined. The resolution of force and displacement measurements can be as low as $\sim 1.0 \mu N$ and ~ 0.2 nm [54] and up to 500mN and $20 \mu m$, respectively [55]. Force actuation is generally obtained using electromagnetic, electrostatic, or piezo-electric means while displacement is determined by capacitance/induction and laser devices [54,55]. Displacement and force measurements are coupled through leaf springs [54]. The low force values and high spatial resolution of nanoindentation systems allow for the testing of small, thin samples and the determination of mechanical properties of sub-microstructural features encountered in biological materials.

The most common properties determined from the force-displacement curves are the stiffness and hardness of the material. Procedures utilized to compute stiffness and hardness include the

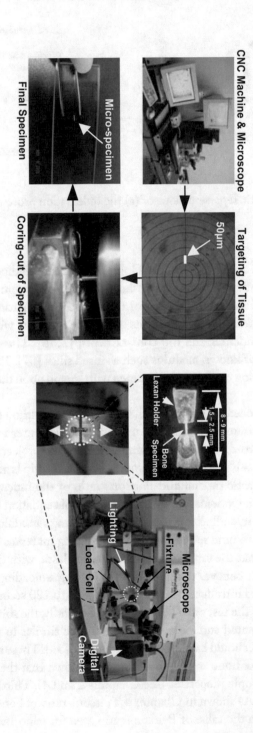

Figure 3.17: (A) Steps in the process of obtaining micro-specimens for tensile testing. (B) Close-up of tensile test specimen, specimen in the fixture, and general view of mechanical testing and photo microscopy system.

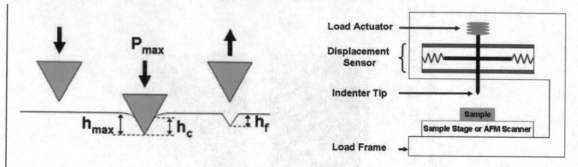

Figure 3.18: Schematic representations of (a) the indentation process and (b) components of a nanoindentation system [1].

Oliver-Pharr method and it variants, the Gong-Maio-Peng method, the Bao-Wang-Zhou method, and the so-called Continuous Stiffness method [56]. The most common method is the Oliver-Pharr, having been utilized in nearly all reports of the stiffness and hardness of bone and teeth [56]. This method requires determining the constants of a shape function, which relates the projected contact area created by the indenter and the contact depth [1, 56]. The constants can be determined by indenting a material of known modulus such as fused silica [57]. The number of calibration indents made and shape function constants determined will depend upon the accuracy that the user is seeking to achieve.

The loading procedure utilized in the Oliver-Pharr method consists of incrementally loading the indenter to a maximum force and then unloading the indenter at the same rate (Figure 3.19). The hardness is defined as the ratio of the maximum load to the projected contact area. When expressed mathematically, the load "P" and displacement "h" relationship is given as $P = B(h - h_f)m$, where B incorporates the elastic moduli and Poisson's ratio of the indenter and test material and m is a parameter related to the indenter tip geometry [56]. This equation can be related to the slope of the upper portion of the unloading curve to determine Young's modulus for the test material.

It is important to note several points when testing materials using the Oliver-Pharr method. The first of these is that the sample is assumed to be elastic with time independent plasticity [58]. That is, the specimen behaves purely elastically during unloading [56]. A means of minimizing viscoelastic effects is to introduce a holding period of 3 to 120 seconds [1] between the loading and unloading portions of the test cycle (Figure 3.19). Secondly, the solution for the elastic deformation of an irreversibly indented surface geometry should be similar to that for a flat semi-infinite half space and the material should have isotropic properties [56]. These requirements may not be met for mineralized tissues like bone and teeth in that it is known that these materials exhibit hierarchical structures and anisotropic properties (see Chapters 2 and 4). Third, the value of Poisson's ratio for the sample is known. As shown in Chapter 4, Poisson's ratio of bone also exhibits anisotropy, which may have a bearing on the value of Poisson's ratio used for nano-indentation calculations.

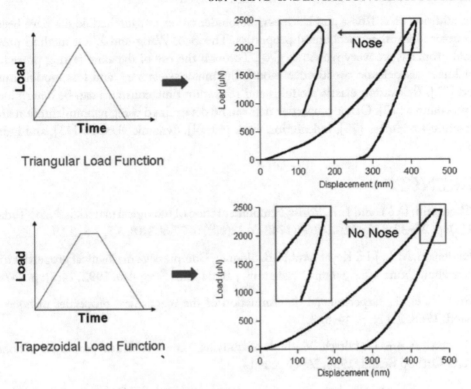

Figure 3.19: Schematic representations of two commonly used loading cycles. The upper figure (A) represents a typical loading-unloading cycle. Curves obtained for viscoelastic materials may exhibit a "nose" which precedes the unloading portion of the curve and may lead to erroneous stiffness values. Viscoelastic effects can be minimized by inserting a holding period between the loading and unloading portions of the cycle (B) [1].

Issues related to system performance and specimen preparation also have a bearing on the use of nanoindentation for determining material properties. Calibration of the indenter is important as the tip radius has been shown to blunt with use [59]. Corrections must also be made for thermal drift and system compliance [60]. The state of specimen hydration has been shown to influence stiffness and hardness values, with dry specimens exhibiting higher stiffness and hardness values than wet [61–64]. Surface roughness has also been shown to influence stiffness and hardness [65], and it has been suggested that indentation depths be such that the surface roughness is less than 10% of the penetration depth [66]. The elastic modulus and viscosity of the specimen embedding media may significantly affect the measurement [67]. Stiffness and hardness can also be affected by the specimen storage media and time of storage [67,68]. Moreover, the stiffness and plastic deformation are dependent on loading rate and load-time cycle [69].

In addition to stiffness and hardness, nanoindentation test methodologies have been developed to determine other mechanical properties. The Bao, Wang and Zhou method provides an additional property, recovery resistance [70]. Through the use of dynamic testing procedures and use of a linear viscoelastic constitutive model, the material's storage and loss moduli can be determined [71]. Relaxation elastic modulus and relaxation time constants can be determined using similar procedures [72]. Other properties that can be determined using nanoindentation are plane-strain fracture toughness [73], indentation work [69, 74], dynamic viscosity [75], and indentation creep [74].

REFERENCES

[1] Ebenstein, D.M. and L.A. Pruitt, Nanoindentation of biological materials. Nano Today, 2006. 1(3): p. 26–33. DOI: 10.1016/S1748-0132(06)70077-9 3.18, 3.5, 3.5, 3.19

[2] Anderson, M.J., J.H. Keyak, and H.B. Skinner, Compressive mechanical properties of human cancellous bone after gamma irradiation. J Bone Joint Surg Am, 1992. 74(5): p. 747–52. 3.1

[3] Bird, F., et al., Experimental determination of the mechanical properties of bone. Aerosp Med, 1968. 39(1): p. 44–8. 3.1

[4] Brodetti, A. and C. Hirsch, Methods of studying some mechanical properties of bone tissue. Acta Orthop Scand, 1956. 26(1): p. 1–14. 3.1

[5] Gravius, S., et al., [Mechanical in vitro testing of fifteen commercial bone cements based on polymethylmethacrylate]. Z Orthop Unfall, 2007. 145(5): p. 579–85. DOI: 10.1055/s-2007-965666 3.1

[6] Grimal, Q., et al., Assessment of cortical bone elasticity and strength: Mechanical testing and ultrasound provide complementary data. Med Eng Phys, 2009. DOI: 10.1016/j.medengphy.2009.07.011 3.1

[7] Linde, F., et al., Mechanical properties of trabecular bone by a non-destructive compression testing approach. Eng Med, 1988. 17(1): p. 23–9. DOI: 10.1243/EMED_JOUR_1988_017_008_02 3.1

[8] Schoenfeld, C.M., E.P. Lautenschlager, and P.R. Meyer, Jr., Mechanical properties of human cancellous bone in the femoral head. Med Biol Eng, 1974. 12(3): p. 313–7. DOI: 10.1007/BF02477797 3.1

[9] Wang, X., et al., The role of collagen in determining bone mechanical properties. J Orthop Res, 2001. 19(6): p. 1021–6. DOI: 10.1016/S0736-0266(01)00047-X 3.1

[10] Boger, A., et al., NMP-modified PMMA bone cement with adapted mechanical and hardening properties for the use in cancellous bone augmentation. J Biomed Mater Res B Appl Biomater, 2009. 90(2): p. 760–6. DOI: 10.1002/jbm.b.31345 3.1

[11] DeTora, M. and K. Kraus, Mechanical testing of 3.5 mm locking and non-locking bone plates. Vet Comp Orthop Traumatol, 2008. 21(4): p. 318–22. 3.1

[12] Goss, B., et al., Elution and mechanical properties of antifungal bone cement. J Arthroplasty, 2007. 22(6): p. 902–8. DOI: 10.1016/j.arth.2006.09.013 3.1

[13] Ignatius, A.A., et al., Composites made of rapidly resorbable ceramics and poly(lactide) show adequate mechanical properties for use as bone substitute materials. J Biomed Mater Res, 2001. 57(1): p. 126–31. DOI: 10.1002/1097-4636(200110)57:1%3C126::AID-JBM1151%3E3.0.CO;2-M 3.1

[14] Lee, A.J., et al., Factors affecting the mechanical and viscoelastic properties of acrylic bone cement. J Mater Sci Mater Med, 2002. 13(8): p. 723–33. DOI: 10.1023/A:1016150403665 3.1

[15] Wang, X.D. and J.S. Nyman, A novel approach to assess post-yield energy dissipation of bone in tension. Journal of Biomechanics, 2007. 40(3): p. 674–677. DOI: 10.1016/j.jbiomech.2006.02.002 3.2.4, 3.3.4, 3.3.4, 3.9

[16] Hoffmeister, B.K., et al., Anisotropy of Young's modulus of human tibial cortical bone. Medical & Biological Engineering & Computing, 2000. 38(3): p. 333–338. DOI: 10.1007/BF02347055 3.3

[17] Swadener, J.G., J.Y. Rho, and G.M. Pharr, Effects of anisotropy on elastic moduli measured by nanoindentation in human tibial cortical bone. Journal of Biomedical Materials Research, 2001. 57(1): p. 108–112. DOI: 10.1002/1097-4636(200110)57:1%3C108::AID-JBM1148%3E3.0.CO;2-6 3.3

[18] Turner, C.H., et al., The elastic properties of trabecular and cortical bone tissues are similar: results from two microscopic measurement techniques. Journal of Biomechanics, 1999. 32(4): p. 437–441. DOI: 10.1016/S0021-9290(98)00177-8 3.3

[19] Yeni, Y.N., et al., Apparent viscoelastic anisotropy as measured from nondestructive oscillatory tests can reflect the presence of a flaw in cortical bone. Journal of Biomedical Materials Research Part A, 2004. 69A(1): p. 124–130. 3.3

[20] Augat, P., et al., Anisotropy of the elastic modulus of trabecular bone specimens from different anatomical locations. Medical Engineering & Physics, 1998. 20(2): p. 124–131. DOI: 10.1016/S1350-4533(98)00001-0 3.3

[21] Birnbaum, K., et al., Material properties of trabecular bone structures. Surgical and Radiologic Anatomy, 2001. 23(6): p. 399–407. DOI: 10.1007/s00276-001-0399-x 3.3

[22] Brunet-Imbault, B., et al., Anisotropy measurement of trabecular bone radiographic images using spectral analysis. Journal of Bone and Mineral Research, 2001. 16: p. S462-S462. 3.3

[23] Tabor, Z. and E. Rokita, Quantifying anisotropy of trabecular bone from gray-level images. Bone, 2007. 40(4): p. 966–972. DOI: 10.1016/j.bone.2006.10.022 3.3

[24] Currey, J.D., Effects of Strain Rate, Reconstruction and Mineral Content on Some Mechanical-Properties of Bovine Bone. Journal of Biomechanics, 1975. 8(1): p. 81-&. DOI: 10.1016/0021-9290(75)90046-9 3.3, 3.3.1.2

[25] Currey, J.D., Strain Rate Dependence of the Mechanical-Properties of Reindeer Antler and the Cumulative Damage Model of Bone-Fracture. Journal of Biomechanics, 1989. 22(5): p. 469–475. DOI: 10.1016/0021-9290(89)90207-8 3.3, 3.3.1.2

[26] Ferreira, F., M.A. Vaz, and J.A. Simoes, Mechanical properties of bovine cortical bone at high strain rate. Materials Characterization, 2006. 57(2): p. 71–79. DOI: 10.1016/j.matchar.2005.11.023 3.3, 3.3.1.2

[27] Hansen, U., et al., The effect of strain rate on the mechanical properties of human cortical bone. Journal of Biomechanical Engineering-Transactions of the Asme, 2008. 130(1): p. -. DOI: 10.1115/1.2838032 3.3, 3.3.1.2

[28] LaMothe, J.M. and R.F. Zernicke, Mechanical loading rate and strain gradients positively relate to periosteal bone formation rate. Faseb Journal, 2004. 18(5): p. A782-a782. 3.3, 3.3.1.2

[29] Linde, F., et al., Mechanical-Properties of Trabecular Bone - Dependency on Strain Rate. Journal of Biomechanics, 1991. 24(9): p. 803–809. DOI: 10.1016/0021-9290(91)90305-7 3.3, 3.3.1.2

[30] Schaffler, M.B., E.L. Radin, and D.B. Burr, Mechanical and Morphological Effects of Strain Rate on Fatigue of Compact-Bone. Bone, 1989. 10(3): p. 207–214. DOI: 10.1016/8756-3282(89)90055-0 3.3, 3.3.1.2

[31] Vanleene, M., P.E. Mazeran, and M.C.H.B. Tho, Influence of strain rate on the mechanical behavior of cortical bone interstitial lamellae at the micrometer scale. Journal of Materials Research, 2006. 21(8): p. 2093–2097. 3.3, 3.3.1.2

[32] ASTM, Standard testing for compressive strength of cylindrical concrete specimens. ASTM Standard C, 1991: p. 39–86. 3.3.1.2

[33] Courtney, A.C., W.C. Hayes, and L.J. Gibson, Age-related differences in post-yield damage in human cortical bone. Experiment and model. J Biomech, 1996. 29(11): p. 1463–71. DOI: 10.1016/0021-9290(96)84542-8 3.3.4

[34] Jepsen, K.J. and D.T. Davy, Comparison of damage accumulation measures in human cortical bone. Journal of Biomechanics, 1997. 30(9): p. 891–4. DOI: 10.1016/S0021-9290(97)00036-5 3.3.4

[35] Jepsen, K.J., D.T. Davy, and D.J. Krzypow, The role of the lamellar interface during torsional yielding of human cortical bone. Journal of Biomechanics, 1999. 32(3): p. 303–10. DOI: 10.1016/S0021-9290(98)00179-1 3.3.4

[36] Nyman, J.S., et al., Mechanical behavior of human cortical bone in cycles of advancing tensile strain for two age groups. J Biomed Mater Res A, 2009. 89(2): p. 521–9. DOI: 10.1002/jbm.a.31974 3.3.4, 3.3.4

[37] Leng, H., X.N. Dong, and X. Wang, Progressive post-yield behavior of human cortical bone in compression for middle-aged and elderly groups. J Biomech, 2009. 42(4): p. 491–7. DOI: 10.1016/j.jbiomech.2008.11.016 3.3.4, 3.3.4

[38] Fondrk, M.T., E.H. Bahniuk, and D.T. Davy, Inelastic strain accumulation in cortical bone during rapid transient tensile loading. J Biomech Eng, 1999. 121(6): p. 616–21. DOI: 10.1115/1.2800862 3.3.4

[39] Schaffler, M.B., E.L. Radin, and D.B. Burr, Long-term fatigue behavior of compact bone at low strain magnitude and rate. Bone, 1990. 11(5): p. 321–6. DOI: 10.1016/8756-3282(90)90087-F 3.3.7

[40] Carter, D.R. and W.E. Caler, A cumulative damage model for bone fracture. Journal of Orthopaedic Research, 1985. 3(1): p. 84–90. DOI: 10.1002/jor.1100030110 3.3.7

[41] Gustafson, M.B., et al., Calcium buffering is required to maintain bone stiffness in saline solution. J Biomech, 1996. 29(9): p. 1191–4. DOI: 10.1016/0021-9290(96)00020-6 3.3.7

[42] Ascenzi, A. and E. Bonucci, The ultimate tensile strength of single osteons. Acta Anat, 1964. 58: p. 160–183. DOI: 10.1159/000142580 3.4

[43] Ascenzi, A. and B. F, The tensile properties of single osteons. Anatomical Record, 1967. 158(4): p. 375–86. DOI: 10.1002/ar.1091580403 3.4

[44] Reyes, M., R. Acuna, and X. Wang. Micromechanical tensile testing of cortical bone tissue. in 53rd Annual Meeting of the Orthopaedics Research Society. 2007. San Diego, CA. 3.4

[45] Ascenzi, A. and E. Bonucci, The shearing properties of single osteons. Anat Rec, 1968. 172: p. 499–510. DOI: 10.1002/ar.1091720304 3.4, 3.4

[46] Hernandez, C.J., et al., Trabecular microfracture and the influence of pyridinium and non-enzymatic glycation-mediated collagen cross-links. Bone, 2005. 37(6): p. 825–32. DOI: 10.1016/j.bone.2005.07.019 3.4

[47] Choi, K., et al., The elastic moduli of human subchondral, trabecular, and cortical bone tissue and the size-dependency of cortical bone modulus. J Biomech, 1990. 23(11): p. 1103–13. DOI: 10.1016/0021-9290(90)90003-L 3.4

[48] Choi, K. and S.A. Goldstein, A comparison of the fatigue behavior of human trabecular and cortical bone tissue. J Biomech, 1992. 25(12): p. 1371–81. DOI: 10.1016/0021-9290(92)90051-2 3.4

[49] Ascenzi, A., B. P, and B. A, The bending properties of single osteons. Journal of Biomechanics, 1990. 23(8): p. 763–71. DOI: 10.1016/0021-9290(90)90023-V 3.4

[50] Reyes, M., et al. Agine affects toughness and ultrastructure of osteonal and interstitial bone. in 52nd Annual Meeting of the Orthopaedic Research Society. 2006. Chicago, IL. 3.4

[51] Ascenzi, A., B. P, and B. A, The torsional properties of single selected osteons. Journal of Biomechanics, 1994. 27(7): p. 875–84. DOI: 10.1016/0021-9290(94)90260-7 3.4

[52] Ascenzi, A. and B. E, The compressive properties of single osteons. Anatomical Record, 1968. 161(3): p. 377–91. DOI: 10.1002/ar.1091610309 3.4

[53] Reyes, M., et al. Age-related changes in the compressive properties of interstitial and osteonal tissue. in 55th Annual Meeting of the Orthopaedic Research Society. 2009. Las Vegas, NV. 3.4

[54] VanLandingham, M., Review of instrumented indentation. Journal of Research of the National Institute of Standards and Technology, 2003. 108(4): p. 249–265. 3.5

[55] Fischer-Cripps, A., Nanoindentation. 2004, Springer. 3.5

[56] Lewis, G. and J.S. Nyman, The use of nanoindentation for characterizing the properties of mineralized hard tissues: state-of-the art review. J Biomed Mater Res B Appl Biomater, 2008. 87(1): p. 286–301. DOI: 10.1002/jbm.b.31092 3.5, 3.5

[57] Oliver, W. and G. Pharr, An improved technique for determining harness and elastic modulus using load and displacement sensing indentation experiments. Journal of Materials Research, 1992. 7: p. 1564–1583. DOI: 10.1557/JMR.1992.1564 3.5

[58] Haque, F., Application of Nanoindentation Development of Biomedical to Materials. Surface Engineering, 2003. 19(4): p. 255. DOI: 10.1179/026708403322499173 3.5

[59] Bull, S., Extracting hardness and Young's modulus from load-displacement curves. Zeitschrift für Metallkunde, 2002. 93(9): p. 870–874. 3.5

[60] Menclk, J. and M.V. Swain, Errors associated with depth-sensing micro-indentation tests. Journal of Materials Research, 1995. 10(6): p. 1491–1501. DOI: 10.1557/JMR.1995.1491 3.5

[61] Hoffler, C.E., et al., An application of nanoindentation technique to measure bone tissue Lamellae properties. J Biomech Eng, 2005. 127(7): p. 1046–53. DOI: 10.1115/1.2073671 3.5

[62] Ammann, P., et al., Strontium ranelate treatment improves trabecular and cortical intrinsic bone tissue quality, a determinant of bone strength. J Bone Miner Res, 2007. 22(9): p. 1419–25. DOI: 10.1359/jbmr.070607 3.5

[63] Rho, J.Y. and G.M. Pharr, Effects of drying on the mechanical properties of bovine femur measured by nanoindentation. J Mater Sci Mater Med, 1999. 10(8): p. 485–8. DOI: 10.1023/A:1008901109705 3.5

[64] Ho, S.P., et al., The effect of sample preparation technique on determination of structure and nanomechanical properties of human cementum hard tissue. Biomaterials, 2004. 25(19): p. 4847–57. DOI: 10.1016/j.biomaterials.2003.11.047 3.5

[65] Xu, J., et al., Atomic force microscopy and nanoindentation characterization of human lamellar bone prepared by microtome sectioning and mechanical polishing technique. J Biomed Mater Res A, 2003. 67(3): p. 719–26. DOI: 10.1002/jbm.a.10109 3.5

[66] Habelitz, S., et al., Mechanical properties of human dental enamel on the nanometre scale. Arch Oral Biol, 2001. 46(2): p. 173–83. DOI: 10.1016/S0003-9969(00)00089-3 3.5

[67] Mittra, E., S. Akella, and Y.X. Qin, The effects of embedding material, loading rate and magnitude, and penetration depth in nanoindentation of trabecular bone. J Biomed Mater Res A, 2006. 79(1): p. 86–93. DOI: 10.1002/jbm.a.30742 3.5

[68] Hairul Nizam, B.R., et al., Nanoindentation study of human premolars subjected to bleaching agent. Journal of Biomechanics, 2005. 38(11): p. 2204–2211. DOI: 10.1016/j.jbiomech.2004.09.023 3.5

[69] Fan, Z. and J.Y. Rho, Effects of viscoelasticity and time-dependent plasticity on nanoindentation measurements of human cortical bone. J Biomed Mater Res A, 2003. 67(1): p. 208–14. DOI: 10.1002/jbm.a.10027 3.5

[70] Bao, Y.W., W. Wang, and Y.C. Zhou, Investigation of the relationship between elastic modulus and hardness based on depth-sensing indentation measurements. Acta Materialia, 2004. 52(18): p. 5397–5404. DOI: 10.1016/j.actamat.2004.08.002 3.5

[71] Odegard, G., T. Gates, and H. Herring, Characterization of viscoelastic properties of polymeric materials through nanoindentation. Experimental Mechanics, 2005. 45: p. 130–136. DOI: 10.1007/BF02428185 3.5

[72] Balooch, M., et al., Viscoelastic properties of demineralized human dentin measured in water with atomic force microscope (AFM)-based indentation. J Biomed Mater Res, 1998. 40(4): p. 539–44. DOI: 10.1002/(SICI)1097-4636(19980615)40:4%3C539::AID-JBM4%3E3.3.CO;2-O 3.5

[73] Lawn, B., A. Evans, and D. Marshall, Elastic/Plastic Indentation Damage in Ceramics: The Median/Radial Crack System. 1980. p. 574–581. 3.5

[74] He, L.H. and M.V. Swain, Energy absorption characterization of human enamel using nanoindentation. J Biomed Mater Res A, 2007. 81(2): p. 484–92. 3.5

[75] Tang, B., A.H. Ngan, and W.W. Lu, An improved method for the measurement of mechanical properties of bone by nanoindentation. J Mater Sci Mater Med, 2007. 18(9): p. 1875–81. DOI: 10.1007/s10856-007-3031-8 3.5

CHAPTER 4

Mechanical Behavior of Bone

CHAPTER SUMMARY

In order to provide tissue engineers with comprehensive information on the mechanical behavior of bone, this chapter presents a broad experimental data base pertaining to the mechanical properties of bones at different length scales (i.e., from whole bone to nano-level constituents). In addition, this chapter presents of an overview of the *in vivo* response of bone to physiologically relevant loading. The chapter is broadly divided into eight sections (4.1–4.8). Section 4.1 provides a brief overview of the hierarchical structure of bone and issues that should be considered when conducting mechanical tests of bone and interpreting the results. Section 4.2 addresses the mechanical properties of whole bone. Section 4.3 will address the mechanical properties of macroscopic cortical and cancellous bone specimens for which continuum assumptions hold. Section 4.4 presents the mechanical properties of micro-structural features that comprise cortical and cancellous bone. Section 4.5 will present lamellar level mechanical properties. Section 4.6 will review the results of mechanical tests that have been conducted on the collagen fibrils and hydroxyapatite crystals. Section 4.7 will return back to the whole bone level with a presentation of the mechanical response of bone to physiologically relevant loading. Section 4.8 provides an overview of the chapter and concluding remarks.

4.1   INTRODUCTION

As presented in Chapter 2, bone has a hierarchical organization of an organic phase (Type 1 collagen and non-collagenous proteins) and a mineral phase with water interspersed throughout the tissue. This hierarchical structure is organized into levels ranging from macro to nanoscopic length scales [1, 2]: (1) whole bone; (2) cortical and cancellous bone at the macrostructural level; (3) Haversian systems (cortical bone) or trabeculae (cancellous bone) at the microstructural level; (4) lamellar tissue at the sub-microstructural level; (5) mineral matrix and collagen fibrils at the nano-structural level; and finally, (6) the individual collagen molecules, mineral crystals, and water molecules at the sub-nanostructural level.

   Due to the hierarchical structure of bone, tests conducted at a given length scale may be influenced by material and structural characteristics of the bone at the same or lower levels of hierarchy. For example, an osteon consists of concentric lamellae that are comprised of directionally oriented collagen fibrils within a mineral matrix (Figure 2.1). The overall mechanical properties of osteons may thereby be affected by a structural factor such as lamellar orientation and a compositional factor such as degree of mineralization. Similar considerations apply to the individual trabeculae that are obtained from cancellous bone. Thus, consideration should be given to the structural and material factors that may influence the interpretation of results reported at a given hierarchical level.

Before proceeding, the reader is directed to *Chapter 2: Bone Composition and Structure* for an overview of the composition and hierarchical structure of bone. *Chapter 3: Current Test Methodologies of Bone* presents an overview of the mechanical tests which have been developed to study the mechanical properties of bone and which are the basis for the information that is presented in this chapter.

4.2   MECHANICAL PROPERTIES OF WHOLE BONE

Whole bone tests can provide useful information on the strength and stiffness of the whole structure under simulated loading conditions. In most cases, biomechanical properties are measured for relevant physiological loading conditions in order to assess the *in vivo* behavior of the whole bones. As shown in Table 4.1, human vertebrae experience mainly compressive load *in vivo*. Thus, compressive tests are the most common tests performed on human vertebrae to study the *in vivo* behavior of these bones. In addition, since most long bones may experience both torsion and bending *in vivo*, the majority of mechanical tests for long bones are performed in three-point / four-point bending or torsional configurations. Moreover, numerical methods, such as imaging-based finite element modeling (FEM), have been frequently used to predict the overall mechanical behavior of skeletal structures to impact loads by simulating the boundary conditions of a fall on the hip. In this case, whole bone tests become necessary to validate the FEM simulations. For bone tissue engineers, the information on the mechanical properties of normal whole bones is very important to help design the appropriate tissue engineered bone implants in the bones.

In general, whole bone tests are structural tests that do not separate the contribution of geometry from that of tissue-level properties to the mechanical behavior of the bone as whole organ. Accounting for structure through the use of the moment of inertia and the flexural equations does not necessarily provide accurate measurements of the material properties of bone tissue. This is largely due to the non-uniform and irregular geometry of whole bones. Due to this limitation, whole bone tests are not suitable for making quantitative comparisons of bone quality between species and even between different bones from the same individuals. Another limitation of *in vitro* whole bone tests is that they are usually performed in simple loading modes, which sometimes may lack biofidelity to *in vivo* loading conditions applied to the bones. These shortcomings necessitate the use of standardized test specimens from the whole bones in order to determine their material properties.

Abundant data are available in the literature regarding the mechanical properties of whole bones from both humans and animals (Tables 4.1 and 4.2). Since the mechanical properties of individual human bones are considerably dependent on age, gender, race, weight, height, daily activities, and other factors, average values are presented in Table 4.1.

These data provide a ballpark estimation of the mechanical behavior of different human bones under various loading conditions. Since the standard material tests of bone samples from small animal models (e.g., rabbit, rat, mice, etc.) are not practical, mechanical properties from whole bone tests of

Table 4.1: Mechanical properties of human whole bone (average values).

Bone	Test	Property	Value	Comments	Ref.
Femur	4pt-Bending	Stiffness (N/mm)	3,250 2,500	Antero-posterior Latero-medial	[17]
	Torsion	Stiffness (N·m/deg)	6.5		[17]
		Stiffness (N·m/rad)	562		[18]
		Max. Torque (N·m)	183		[18]
	Fall Simulation	Strength (N)	3,934		[19]
Tibia	4pt-Bending	Stiffness (N/mm)	1,900 1,270	Antero-posterior Latero-medial	[20]
	Torsion	Stiffness (N·m/deg)	2.6		[20]
		Stiffness (N·m/rad)	326		[18]
		Max. Torque (N·m)	101		[18]
Humerus	3pt-Bending	Stiffness (N/mm)	285.5 259	Antero-posterior latero-medial	[21]
	Torsion	Stiffness (N·m/deg)	2.53		[21]
		Failure Torque (N·m)	45.6		[21]
Vertebrae	Compression	Max. Nominal Stress (MPa)	2.6 to 6.0	$T_5 - T_7$, L_3 20 to 80 year olds	[22]
			0.2 to 4.2	L_3 43 to 95 year olds (average 81)	[23]
			5.70 5.83	Female Male; < 50 years old	[24]
			2.62 2.96	Female Male; 50 – 75 years old	[24]
			2.29 1.95	Female Male; > 75 years old	[24]

small animal bones are provided in Table 4.2. Although only relative comparisons can be made, these data of normal bones listed in the table can serve as controls for evaluating the tissue engineered bone samples from the animal models.

Animal models have served as valuable tools for *in vitro* and *in vivo* studies on the mechanical properties of bone, pathological bone conditions [3–5], and the *in vivo* properties of tissue engineering products [6–9]. Bones from large animals, such as femurs and tibias from cows, are popular tissues in biomechanical studies due to their large size and ready availability. Small animals, such as genetically altered mice, have been developed to mimic pathological conditions such as osteogenesis imperfecta [10–12] and allow for investigations of the role of structural alterations on the mechanical behavior of bone.

Table 4.2: Mechanical properties of animal whole bone (average values). *(Continues.)*

Species	Bone	Test	Property	Value	Comment	Ref.
Bear	Femur	3pt-Bending	Ultimate Stress (MPa)	238	Spring Bears	[25]
			Failure Energy (J)	46	Spring Bears	[25]
			Modulus of Toughness (mJ/mm^3)	6.1	Spring Bears	[25]
Monkey	Tibia	3pt-Bending	Stiffness (MPa)	9,044		[26]
		Torsion	Failure Torque (N mm)	3,782		[26]
			Shear Modulus (MPa)	2,849		[26]
			Max. Shear Stress (MPa)	47		[26]
Rabbit	Femur	3pt-Bending	Stiffness (N/mm)	288		[27]
			Strength (N)	484		[27]
	Tibia	3pt-Bending	Stiffness (N/mm)	193		[27]
			Strength (N)	413		[27]
	Humerus Femur Tibia	3pt-Bending	Stiffness (MPa)	3,813 3,084 2,714		[28]
	Vertebrae	Compression	Stiffness (MPa)	67,426~76,190	L3 – L7	[29]
			Strength (N)	1736~1862	L3 – L7	[29]
Canine	Femur	Compression	Ultimate Load (N)	5,270		[30]
			Stiffness (N/mm)	5,490		[30]
	Metacarpal	4pt-Bending	Stiffness (kN/mm)	0.83		[31]
			Max. Moment (N m)	7.8		[31]
			Energy Absorbed (J)	2		[31]
		Torsion	Stiffness (N m/deg)	0.09		[31]
			Max. Torque (Nm)	1.25		[31]
			Energy Absorbed (J)	0.25		[31]
Cat	Femur	3pt-Bending	Force to Elastic Limit (N)	331.6		[32]

Table 4.2: *(Continued.)* Mechanical properties of animal whole bone (average values). *(Continues.)*

			Force to Failure (N)	441.9		[32]
			Stiffness (N/mm)	264.4		[32]
			Resilience (mJ)	224		[32]
			Energy to Failure (mJ)	609		[32]
			Stiffness (N/mm)	265		[27]
			Strength (N)	442		[27]
	Tibia	3pt-Bending	Force to Elastic Limit (N)	482.8		[32]
			Force to Failure (N)	531.3		[32]
			Stiffness (N/mm)	207.5		[32]
			Resilience (mJ)	570		[32]
			Energy to Failure (mJ)	863		[32]
			Stiffness (N/mm)	208		[27]
			Strength (N)	531		[27]
Chipmunk	Femur Tibia Humerus	3pt-Bending	Ultimate Stress (MPa)	263.2 303.9 192.5		[33]
Ground Squirrel	Femur Tibia Humerus	3pt-Bending	Ultimate Stress (MPa)	219.0 184.6 171.8		[33]
Kagaroo Rat	Femur Tibia	3pt-Bending	Ultimate Stress (MPa)	210.7 218.6		[33]
Painted Quail	Femur Tibia Metatarsus	3pt-Bending	Ultimate Stress (MPa)	311.0 170.5 245.2		[33]
Bobwhite	Femur Tibia Metatarsus	3pt-Bending	Ultimate Stress (MPa)	193.3 294.0 226.3		[33]
Avian	Humerus Ulna Radius Femur Tibiotarsus, Tarsometatarsus	3pt-Bending	Bending Strength (MPa)	167.35 191.05 242.00 145.65 227.08 181.0	29 species	[34]

Table 4.2: *(Continued.)* Mechanical properties of animal whole bone (average values). *(Continues.)*

	Humerus			10.49		
	Ulna			12.06		
	Radius	3pt-Bending	Stiffness (GPa)	20.98	29 species	[34]
	Femur			9.69		
	Tibiotarsus,			16.63		
	Tarsometatarsus			11.96		
Rat	Femur	3pt-Bending	Ultimate Stress (MPa)	253.1		[33]
	Tibia			233.1		
	Humerus			257.5		
	Femur	3pt-Bending	Stiffness (GPa)	4.9 – 6.9		[35, 36]
			Stiffness (N/mm)	113		[27]
			Strength (N)	80.5		[27]
			Stiffness (N/mm)	230	15 day old	[37]
			Yield Load (N)	90.2	15 day old	[37]
			Post-yield Load (N)	46.7	15 day old	[37]
			Ultimate Load (N)	136.9	15 day old	[37]
		Torsion	Stiffness x 10^2 (N m/deg)	8, 8	14, 17 months old	[38]
			Strength x 10^2 (N m)	68,78	14, 17 months old	[38]
			Energy Absorption (N m x deg)	3.3, 4.5	14, 17 months old	[38]
	Tibia	3pt-Bending	Stiffness (GPa)	12.1		[36]
			Stiffness (N/mm)	62.5		[27]
			Strength (N)	48.9		[27]
Mouse	Femur	3pt-Bending	Stiffness (GPa)	8.8		[39]
			Breaking Force (N)	19.9		[40]
			Stiffness (N/mm)	70.0		[40]
			Ultimate Stress (MPa)	123		[40]
			Elastic Modulus (GPa)	1.92		[40]
			Stiffness (N/mm)	56.7		[27]
			Strength (N)	11.4		[27]
			Stiffness (N/mm)	274.3 495.6	Male, Female	[41]

Table 4.2: *(Continued.)* Mechanical properties of animal whole bone (average values).

	4pt-Bending	Work to Yield (mJ)	1.06 1.02	Male, Female	[41]
		Work to Failure (mJ)	2.63 2.75	Male, Female	[41]
		Ult. Energy (N*rad/mm)	0.51	24 weeks old	[42]
	Cantilever Bending	Breaking Moment (g*mm)	2141.5		[43]
	Torsion	Ult. Energy (N mm/mm^2)	0.68	24 weeks old	[42]
Tibia	3pt-Bending	Breaking Force (N)	21.1		[40]
		Stiffness (N/mm)	72.9		[40]
		Ultimate Stress (MPa)	224		[40]
		Elastic Modulus (GPa)	3.75		[40]
		Stiffness (N/mm)	51.1		[27]
		Strength (N)	10.7		[27]
Femur Tibia Humerus	3pt-Bending	Stiffness (N/mm)	89.9 56.5 64.7	52 days old	[44]
		Fracture Force (N)	16.5 9.3 8.0	52 days old	[44]
		Fracture Energy (mJ)	3.39 1.48 1.65	52 days old	[44]
		Stiffness (N/mm)	605 523	4 12 months old	[45]
Vertebrae	Compression	Yield force (N)	91.7, 78.5	4, 12 months old	[45]
		Ultimate force (N)	98.7, 80.5	4, 12 months old	[45]
		Ultimate Displacement. (mm)	0.27, 0.26	4, 12 months old	[45]
		Ultimate Energy (N/mm)	9.2, 8.5	4, 12 months old	[45]

However, several issues should be noted regarding the use of data reported from animal models. First, the structure and composition of animal bones vary with species as well as developmental stage [13] [2, 14–16]. Use of data from animal studies should therefore focus more upon the effects of changes in structure and composition on mechanical properties as opposed to using the data as a substitute for the properties of human bones. Additionally, the size of the animal places a serious restriction on the type of mechanical test that can be performed and thereby information that can be obtained [13]. Large animals lend themselves to testing of both whole bone and standardized test specimens, yielding both structural and material information. However, small animals may only allow for whole bone tests, thereby limiting information to structural properties. Further, the animal model selected may be required to be a close analogue to human bone in terms of anatomy, healing response, or response to physiological loading. These factors are evident when observing the wide range of differences between the mechanical properties of human and animal bones presented in Tables 4.1 and 4.2, as well as the material which follows.

4.3   MECHANICAL PROPERTIES OF BONE AT THE MACROSCOPIC LEVEL

Mechanical testing of cortical and cancellous bone at the macroscopic level requires specimens that are of sufficient size to satisfy material continuum conditions. Specimens of cortical bone are usually on the order of several millimeters and typically reflect the structural anisotropy that is present in the whole bone from which they are prepared. Features to be found in macro-specimens of cortical bone will include osteons, interstitial regions, cement lines, resorption spaces, Haversian and Volkmann's canals, lacunae, and canaliculi. If the bone has not undergone significant remodeling, features will typically be primary bone and osteons. Macroscopic specimens of trabecular bone should also be of sufficient size (on the order of 5~10mm) to satisfy continuum conditions [46, 47]. The trabecular network at this level may exhibit an orthogonal organization. In both cortical and cancellous bone macro-specimens, the anisotropy of the micro-structure tends to lead to anisotropic macro-mechanical properties.

4.3.1   ELASTIC PROPERTIES OF CORTICAL AND TRABECULAR MACROSCOPIC BONE

It is well accepted that the structure of cortical bone is transversely isotropic, with osteons being generally oriented along the long axis of the bone [48–50]. As shown in the transverse cross-section of cortical bone (Figure 2.2), osteons are oriented along the long axis and in a random distribution with respect to the transverse plane. The mechanical properties of cortical bones depend on the orientation of loading with respect to the orientation of the osteons. To fully characterize the mechanical properties of cortical bone, mechanical tests must be performed on multiple specimens with varying orientations of the osteons. Due to this natural transverse isotropy, a complete characterization of the elastic behavior of cortical bone usually requires five elastic constants.

The macro-elastic properties of human long bones have demonstrated a dependence on the following factors: anatomic location, loading mode, orientation, degree of mineralization, and specimen size. As to the effect of anatomic locations, the elastic modulus of human tibial bone is higher than that of the femoral counterparts in both tension [51, 52] and compression [52]. The tensile elastic moduli in the longitudinal orientation are similar amongst fibula, radius, and ulna, but higher when compared to the humerus [51]. The tensile elastic moduli of these four types of long bones fall within the range of values reported for femoral specimens.

The macro-elastic properties of human cortical bone are also dependent on loading modes. It has been observed that the elastic modulus of tibial cortical bone is significantly lower in tension than in compression [51, 52]. In addition, the flexural moduli obtained from three-point and four-point bending tests are significantly different from those of uniaxial tension tests [53–56]. It should be noted that unlike a uniaxial tension test, the modulus value obtained from three-point and four-point bending tests is based upon linear elastic beam theory, which may not reflect the true behavior of bone.

In addition, the macro-elastic properties of human cortical bone are dependent on orientation. The elastic modulus of human cortical bone is much lower in the transverse orientation than the longitudinal orientation irrespective of loading modes (i.e., tension and compression) The elastic moduli of bone in radial and lateral orientations (both transverse to the long axis of bone) are similar, but significantly lower than that in the longitudinal orientation [57, 58]. These data verify the transverse isotropic nature of human cortical bones.

It is also noteworthy that the macro-elastic properties of human bone are age-dependent. This is important for tissue engineers to take into account the effects of aging in their design and research of bone tissue engineering products. Age-related decreases in the elastic modulus of femoral cortical bone range between -1.5% and -2.2% per decade in tension and compression, respectively [52]. Age-related changes also occur in the tibia. Unlike the femoral bone, the tibia exhibits an age-related increase in elastic modulus of 1.5% per decade [52]. However, tibiae experience greater age-related decreases in elastic modulus (-4.7% and -4.0% per decade, respectively), similar to their femoral counterparts [52].

Information regarding the yield stress, strain, and Poisson's ratio of cortical bone macro specimens is not as complete as that for stiffness. Longitudinal femoral tension specimens have exhibited yield stresses lower than tibial specimens [52]. Additionally, both femoral and tibial tissues exhibited age-related decreases in yield strength of -2.2% and -0.5% per decade, respectively [52]. These differences in mechanical properties further underscore the differences that may exist between anatomic locations. Poisson's ratio for human cortical bone has been limited to femoral specimens. Reported values for pooled tension and compression specimens have demonstrated anisotropic behavior with values of 0.46 for the response to a transverse load and 0.58 for the response to a longitudinal load [57].

Trabecular bone exhibits a wide variation in elastic properties between bones, anatomic locations within the same bones, and species (Tables 4.6 and 4.7). For example, test specimens from

Table 4.3: Elastic modulus of macroscopic cortical bone.

Species	Bone	Test	Orientation	Modulus (GPa)	Comments	Ref.
Human	Femur	Tension	Longitudinal	15.6 – 19.4		[51, 52, 59]
		Compression	Longitudinal	15.2 – 18.1		[52, 60]
		Tension / Compression	Longitudinal	17	Specimens pooled	[57, 61]
		Tension / compression	Transverse	11.5	Specimens pooled	[57]
		Torsion	Longitudinal	3.3 – 5.0		[57, 62]
		3pt-Bending	Longitudinal	10.8 – 15.8		[53-56]
		4pt-Bending	Longitudinal	12.1		[63]
	Tibia	Tension	Longitudinal	18.0 – 29.2		[51, 52]
		Compression	Longitudinal	25.9 – 35.3		[52]
		Cantilever bending	Longitudinal	10.6		[64]
	Femur, Tibia	Torsion	Longitudinal	3.17 – 3.58	Specimens pooled	[52]
	Fibula	Tension	Longitudinal	18.5		[51]
	Humerus	Tension	Longitudinal	17.2		[51]
	Radius	Tension	Longitudinal	18.5		[51]
	Ulna	Tension	Longitudinal	18.4		[51]
	Femur/ tibia/ humerus	Tension	Longitudinal	11.9	Specimens pooled	[58]
		Compression	Longitudinal	14.2		[58]
		Compression	Radial	3.8		[58]
		Compression	Lateral	4.2		[58]
Bovine	Femur	Tension	Longitudinal	28.9	Laminar tissue	[61]
		Tension	Longitudinal	23.9	Haversian tissue	[61]
		Compression	Longitudinal	18.6	Tissue type not reported	[60]
		Tension / compression	Longitudinal	22.6	Haversian tissue; Specimens pooled	[57]
		Tension / compression	Transverse	10.0	Haversian tissue; Specimens pooled	[57]
		Torsion	Longitudinal	3.6	Haversian tissue; Specimens pooled	[57]
	Tibia	Tension	Longitudinal	21.2 – 28.2		[61, 65, 66]
		Compression	Longitudinal	7.1		[65]
		Bending	Longitudinal	14.1		[65]
		3pt-Bending	Longitudinal	11.0 / 18.6 / 21.0	Haversian-Plexiform / Haversian / Plexiform	[67]
Horse	Femur	4pt-Bending	Longitudinal	16.8 – 19.9	4 quadrants tested	[68]
	Radius	4pt-Bending	Longitudinal	16.2 – 20.2	4 quadrants tested	[68]
Rabbit	Femur	4pt-Bending	Longitudinal	14 – 15		[69]
	Tibia	4pt-Bending	Longitudinal	21 – 22		[69]
	Humerus	4pt-Bending	Longitudinal	13 – 14		[69]
Canine	Femur	3pt-Bending	Longitudinal	3.5		[70]
Equine	3rd Metacarpal	4pt-Bending	Longitudinal	18.3		[71]
Baboon	Femur	Tension	Longitudinal	4.6 – 5.0		[72]
	Femur	3pt-Bending	Longitudinal	9.5 – 13.4		[73]

Table 4.4: Yield stress and strain of cortical bone.

Species	Bone	Test	Yield Stress (MPa)	Yield Strain	Comments	Ref.
Human	Femur	Tension	120 104 -2.2% per decade		Ages 20 to 89 y.o.	[52]
		Tension / compression	114 ± 3.1 (Tension)		Specimens pooled	[57]
		Torsion	55.8 ± 3.8	0.0013±0.001		[62]
		3pt-Bending	154 ± 13			[55]
		3pt-Bending	166 ± 12			[56]
	Tibia	Tension	126 131 -0.5% per decade		Ages 20 to 89 y.o.	[52]
Bovine	Femur	Tension	132 ± 10.6		Haversian and Haversian/Plexiform	[74]
	Femur	Compression	196 ± 18.5		Haversian and Haversian/Plexiform	[74]
	Femur	Torsion	57 ± 8.4		Haversian and Haversian/Plexiform	[74]
	Tibia	Tension	160 ± 15.1	0.0057±0.00116	Haversian	[66]
Horse	Femur	4pt-Bending	97 – 148		4 quadrants	[68]
	Radius	4pt-Bending	89 – 125		4 quadrants	[68]
Baboon	Femur	3pt-Bending	473 – 640			[73]
	Femur	Tension	139 – 154			[72]

Note: All specimens oriented in the longitudinal direction.

Table 4.5: Poisson's ratio for human and animal cortical bone.

Species	Bone	Test	Value	Comments	Ref.
Human	Femur	Tension / compression	0.46	Response to transverse load	[57]
		Tension / compression	0.58	Ratio in isotropic plane	[57]
Bovine	Femur	Tension / compression	0.36	Haversian tissue; Response to transverse load	[57]
		Tension / compression	0.51	Haversian tissue; Ratio in isotropic plane	[57]

Table 4.6: Elastic behavior of trabecular bone. *(Continues.)*

Species	Bone	Test	Specimen Orientation	Stiffness (MPa)	Yield Stress (MPa)	Ref.
Human	Vertebral Body	Tension	SI	139 - 472	1.75 ± 0.65	[75]
		Compression	SI	90 - 536	1.92 ± 0.84	[75]
		Compression	SI	.05 – 462		[86]
		Compression	V	67		[82]
		Compression	H	20		[82]
	Femur	Compression	SI	389.3 ± 270.1		[87]
	Femoral Head	Compression	SI	900 ± 710		[76]
	Femoral Neck	Compression	SI	616 ± 707		[76]
	Femoral Intertrochanter	Compression	SI	263 ± 170		[76]
	Tibial plateau	Compression	L	56.6 ± 9.7		[88]
Canine	Femoral Head	Compression	SI	428 ± 237		[79]
	Femoral Head & Neck	Compression	PD	435	10.7	[77]
	Femur – Distal	Compression	SI, AP, ML	158 – 264		[78]
	Femoral Condyle	Compression	SI	279 – 394		[79]
	Tibial Plateau	Compression	SI	106 – 426		[79]
	Humerus	Compression	L	1490 ± 300	12.89 ± 2.97	[89]
	Humeral Head	Compression	SI	350 ± 171		[79]
	Vertebral Body	Compression	CC	530 ± 40		[90]
Bovine	Femoral Condyle	Compression	L	117.49 ± 61.53		[91]
	Vertebral Body	Compression	CC	173		[92]
Porcine	Vertebral Body	Compression	Unknown	1080 ± 470		[93]
Sheep	Vertebral Body	Compression	CC	1510 ± 784		[94]
	Femoral Neck	Compression	SI	2.004 ± 0.237		[95]

Table 4.6: *(Continued.)* Elastic behavior of trabecular bone.

Monkey	Femoral Head	Compression	PD	372 ± 54	[26]
	Tibia Section	Torsion	N/A	2849 ± 459	[26]

SI: Superior-Inferior; V: Vertical; H: Horizontal; L: Longitudinal; T: Transverse; ML: Medio-Lateral; PD: Proximal-Distal; AP: Anterior-Posterior; CC: Cranial-Caudal

Table 4.7: Poisson's ratio for human and animal trabecular bone.

Species	Bone	Test	Value	Comments	Reference
Bovine	Vertebrae	Compression	0.242 ± 0.099	Drained	[96]
	Vertebrae	Compression	0.399 ± 0.083	Undrained	[96]
Canine	Femur	Ultrasonic	0.14 – 0.32		[97]

vertebrae tested in tension and compression have exhibited approximately three and six-fold differences in modulus values, respectively [75]. Similar behavior has been observed in the femur, where very large standard deviations are evident within each region for specimens taken from the head, neck, and intertrochanteric region [76]. Average stiffness values have also been shown to vary widely for different regions from the same bone. For example, average stiffness values increase when moving from the femoral head to the neck to the intertrochanteric region [76]. Similar behavior is observed in canines between trabecular bone taken from the femoral head and neck [77] when compared to distal tissue [78]. Significant variations are also evident between species. For example, the trabecular bone from the femoral head of humans [76] is over twice as stiff as that obtained from canines [79].

Trabecular bone has also shown a degree of anisotropy in which the structure of the tissue aligns in response to functional loading [80]. Studies have shown trabecular bone to exhibit orthotropic symmetry [47] and transverse isotropy [81]. When the trabeculae are vertically oriented with loading axis, vertebral specimens have three times greater stiffness than when they are horizontally oriented with loading axis [82]. Anisotropy has also been exhibited in bovine bone, in which significant differences are evident in an elastic modulus when compared to the moduli of the orthogonal directions [83].

Apparent density (the mass of bone tissue divided by the bulk volume of the test specimen, including mineralized bone and marrow spaces) significantly influences stiffness. Compressive modulus has been shown to vary as a power law function of apparent density with an exponent of 2 to 3 [84] [85]. These studies show that small changes in apparent density can lead to large changes in compressive moduli. However, the wide range of stiffness values evident for a given apparent density implies the role of other factors in the elastic properties of trabecular bone.

4.3.2   POST-YIELD PROPERTIES OF CORTICAL AND TRABECULAR MACROSCOPIC BONE

Table 4.8 illustrates the post-yield macroscopic properties of cortical bones from humans and animals determined using monotonic tests. Though the influence of anatomy is evident in the post-yield properties of human long bones, it is not as pronounced as the elastic properties. Femoral [51, 52, 57, 59] and tibial [51, 52] bones exhibit the same upper range of strength values at macroscopic levels. However, the lower end in the range of strengths for femoral bone specimens is lower than that of tibial tissue. Tensile strengths of specimens obtained from the fibula, radius, and ulna [51] fall within the middle range of values of femoral specimens. However, humerus tensile strength is near the lower end of the femoral range [51]. The compressive strength of longitudinal specimens is similar for femoral [51, 52, 57, 60] and tibial [51, 52] specimens. However, compression specimens from the fibula, humerus, radius, and ulna [51] all exhibited strengths lower than those observed for the femur and tibia, which experience more load than the bones of the upper extremities.

The influence of specimen orientation and aging are also evident in the post-yield macroscopic behavior of bone. Femoral and tibial tensile specimens oriented in the transverse direction were significantly weaker than those in the longitudinal directions [57]. Age-related differences are evident in the tensile and compressive strengths of femoral and tibial specimens. Femoral and tibial tensile strengths have been demonstrated to decrease at rates of -2.1% and -1.2% per decade, respectively [52]. Compressive strengths have also been shown to decrease for the femur and tibia at rates of -2.5% and -2.0% per decade, respectively [52].

Significant modulus loss with post-yielding is observed in the progressive mechanical tests of human femoral and tibial bones, showing a modulus drop of up to 34% at a maximum applied strain level of 1% [98]. In fact, modulus degradation is a means of quantifying microdamage accumulation in bone. Changes in elastic, yield, viscous, and failure properties are observed to correlate with the microdamage accumulation when human femoral bone specimens are subjected to torsional relaxation cycles with pre-determined degrees of twist [62]. Tensile tests of human femoral specimens utilizing a similar protocol with a post-yield recovery time delay between cycles have also shown the degradation in elastic modulus, secant modulus, stress relaxation and strain recovery [99]. In general, the degradation of mechanical properties of bone after yielding is strain-controlled in both tension and compression [100, 101].

As with its elastic properties, a wide variation in post-yield macroscopic properties is observed for trabecular bone between anatomic locations, between locations within the same bone, and between species (Table 4.9). For example, vertebral specimens tested in compression have demonstrated a wide range of strength values [75, 86]. The femur has also exhibited wide variations in strength with very large standard deviations evident within each region for specimens taken from the head, neck, and intertrochanter [76]. Average strength values can also exhibit significant differ-

Table 4.8: Post-yield macroscopic properties of cortical bone. *(Continues.)*

Species	Bone	Test	Specimen Orientation	Strength (MPa)	Ultimate Strain %	Comm.	Ref.
Human	Femur	Tension	Longitudinal	120 – 161	2.2 – 3.4		[51, 52, 57, 59]
		Tension	Transverse	53	0.7		[57]
		Compression	Longitudinal	150.3 – 209	1.7		[51, 52, 57, 60], [60]
		Compression	Transverse	131			[57]
		Torsion	Longitudinal	74.1	5.2		[62]
		3pt-Bending	Longitudinal	183.4 - 194			[53] [54, 55]
		4pt-Bending	Longitudinal	174		3 males and 2 females	[63]
	Tibia	Tension	Longitudinal	140.3 – 161	2.3 – 4.0		[51, 52], [52]
		Compression	Longitudinal	158.9 – 213			[51, 52]
	Femur, tibia, humerus	Tension	Longitudinal	78.8		Specimens pooled	[58]
		Compression	Longitudinal	108.8		Specimens pooled	[58]
	Fibula	Tension	Longitudinal	146.1			[51]
		Compression	Longitudinal	122.6			[51]
	Humerus	Tension	Longitudinal	122.6			[51]
		Compression	Longitudinal	132.4			[51]
	Radius	Tension	Longitudinal	149.1			[51]
		Compression	Longitudinal	114.8			[51]
	Ulna	Tension	Longitudinal	148.1			[51]
		Compression	Longitudinal	117.7			[51]
Bovine	Femur	Tension	Longitudinal	162	4.9	Havers. & Havers./Pl exi.	[74]
		Compression	Longitudinal	175.8 – 217	1.9 – 3.3	Havers. & Havers./Pl exi.	[60] [74]
		Torsion	Longitudinal	76	22.4 deg.	Havers. & Havers./Pl exi.	[74]
	Tibia	Tension	Longitudinal	188	3.2	Haverisian Plexi.	[66]
		3pt-Bending	Longitudinal	230.5 / 217.1 / 223.8	1.6 / 2.0 / 1.7	Havers. / Havers.- Plexi.	[67]

Table 4.8: *(Continued.)* Post-yield macroscopic properties of cortical bone.

Horse	Femur	4pt-Bending	Longitudinal	201 – 247	All quadrants tested	[68]
	Radius	4pt-Bending	Longitudinal	217 – 255	All quadrants tested	[68]
	3rd Metacarpal	4pt-Bending	Longitudinal	193.8		[71]
Rabbit	Femur	4pt-Bending	Longitudinal	130 – 137		[69]
	Tibia	4pt-Bending	Longitudinal	195 – 198		[69]
	Humerus	4pt-Bending	Longitudinal	165 – 167		[69]
Canine	Femur	Tension	Longitudinal	156		[70]
Baboon	Femur	Tension	Longitudinal	164 – 190		[72]
Baboon	Femur	3pt-Bending	Longitudinal	584 – 640		[73]

ences between regions of the same bone. Strength values have been found to increase when moving from the femoral head to the neck to the intertrochanteric region [76]. Canine tissue exhibits similar behavior between tissue from the femoral head and neck [77] when compared to distal tissue [78]. Strength differences are also apparent between species. For example, the trabecular bone from the femoral head of canines [79] is three times stronger than that from humans [76].

Post-yield properties of trabecular bone are also influenced by specimen orientation. Tissue oriented in the superior-inferior direction is approximately three times stronger than transversely oriented tissue [82]. A separate study found an age-related increase in the anisotropy index (ratio of vertical to horizontal compressive strength) from 2.0 to 3.2 as donor aged from 20 to 80 years. Further, trabecular bone has demonstrated anisotropy ratios that are loading mode dependent. Anisotropy ratios have been shown to be highest in compression loading, followed by tension and shear [102]. The observed anisotropies are consistent with Wolff's law in that trabecular bone is actively remodeled to deposit tissue at the locations necessary to resist the imposed physiologic loads.

Apparent density has been shown to have a significant influence upon strength. Strength varies as a power law function of apparent density with an exponent approximately equal to 2 [85]. Furthermore, the anisotropy indices for loading modes are also influenced by apparent density. The anisotropy index for compression loading decreases significantly with increasing apparent density while the indices for tension and shear loading remain relatively constant [102]. The wide range of stiffness values that are evident for a given apparent density implicate other factors as contributors to the post-yield properties of trabecular bone.

Table 4.9: Post-yield macroscopic properties of trabecular bone. *(Continues.)*

Species	Bone	Test	Specimen Orientation	Strength (MPa)	Ultimate Strain	Ref.
Human	Vertebral Body	Compression	V & H	4.4 – 1.1		[22]
		Tension	SI	2.23 ± 0.76	0.0159 ± 0.0033	[75]
		Compression	SI	2.23 ± 0.95	0.0145 ± 0.0033	[75]
		Compression	SI	0.001 – 2.87		[86]
		Compression	V	2.45	0.074	[82]
		Compression	H	0.88	0.085	[82]
		Compression	SI	2.05 ± 0.16		[103]
		Compression	ML	0.83 ± 0.12		[103]
		Compression	AP	0.72 ± 0.10		[103]
	Femur	Compression	SI	7.36 ± 4.00		[87]
	Femoral Head	Compression	SI	9.3 ± 4.5		[76]
	Femoral Neck	Compression	SI	6.6 ± 6.3		[76]
	Femoral Intertrochanter	Compression	SI	3.6 ± 2.3		[76]
	Tibial plateau	Compression	L	4.2 ± 0.6		[88]
Canine	Femoral Head	Compression	SI	29 ± 4		[79]
	Femoral Head & Neck	Compression	PD	12.1	0.0288	[77]
	Femur – Distal	Compression	SI, AP, ML	7.12 ± 4.6		[78]
	Femoral Condyle	Compression	SI	14 – 28		[79]
	Tibial Plateau	Compression	SI	5 – 24		[79]
	Humerus	Compression	L	13.07 ± 3.09	0.1399 ± 0.0447	[89]
	Humeral Head	Compression	SI	18 ± 6		[79]
	Vertebral Body	Compression	CC	10.1 ± 2.5		[90]
Bovine	Femoral Condyle	Compression	L	8.52 ± 4.24		[91]
	Vertebrae	Compression	CC	10.6		[92]
	Vertebral Body	Compression	CC	7.1		[92]
	Proximal Humerus	Tension	L	7.6 ± 2.2		[104]

Table 4.9: *(Continued.)* Post-yield macroscopic properties of trabecular bone.

	Proximal Humerus	Compress	L	12.4 ± 3.2		[104]
Porcine	Vertebral Body	Compression	CC	27.5 ± 3.4		[93]
Sheep	Vertebral Body	Compression	CC	22.3 ± 7.06	0.032 ± 0.008	[94]
	Femoral Neck	Compression	SI	3.150 ± 0.337	0.04350 ± 0.00306	[95]
Monkey	Femoral Head	Compression	PD	23.1 ± 5.4		[26]
	Tibia - Shaft Section	Torsion	N/A	47 ± 8		[26]

SI: Superior-Inferior; V: Vertical; H: Horizontal; L: Longitudinal; ML: Medio-Lateral; PD: Proximal-Distal; AP: Anterior-Posterior; CC: Cranial-Caudal

4.3.3   FRACTURE TOUGHNESS OF CORTICAL AND TRABECULAR MACROSCOPIC BONE

In general, fracture of bone consists of two subsequent processes: crack formation and propagation. Thus, the fracture behavior of bone is often assessed using fracture toughness tests. The critical stress intensity factor (K_c) and critical strain energy release rate (G_c) are two major parameters used to assess the fracture toughness of bone. The test methodologies for determining the fracture toughness of bone are presented in Chapter 3.

Due to the transverse anisotropy of bone, the fracture toughness of bone is also dependent on the orientation of cracks. To address this issue, one may orient the initial crack either longitudinally or transversely to the longitudinal axis of bone specimens for fracture toughness tests. In longitudinally oriented specimens, crack propagation occurs along the long axis of the bone, i.e., in a direction parallel to the orientation of osteons. In transversely oriented specimens, crack propagates perpendicular to the long axis of the bone. The fracture toughness in the longitudinal direction is much lower than that in the transverse direction [105,106] (Tables 4.10a,b,c). The rising R-curve behavior of bone in the transverse direction indicates that the toughness of bone increases as cracks propagate in that direction [107,108].

Loading mode also has an influence on the fracture toughness of cortical bone. For both longitudinal and transverse fractures, the fracture toughness in opening mode (Mode I) has been shown to be significantly lower than those obtained in the shear and tear loading modes (Mode II and III) [109,110].

Microstructural features have been shown to influence the fracture toughness of cortical bone. Human cortical bone with fewer and smaller osteons tend to be more susceptible to fracture [111]. Clinically, patients with femoral neck fractures have larger Haversian canals, larger osteons, higher porosity, and fewer osteons per unit area [112,113]. Additionally, testing canine bone has demonstrated that smaller and more numerous osteons prevent catastrophic failure due to slow crack

Table 4.10: (a) Fracture toughness G_c of cortical bone.

Species	Bone	Test	Orientation	K_c (MPa m$^{0.5}$)	Reference
Human	Enamel	CT	Longitudinal	2.07 ± 0.22	[125]
	Dentin	CT	Longitudinal	3.132 ± 0.194	[126]
	Femur	SENB	Transverse	5.0 – 6.4	[127]
	Tibia	CT	Longitudinal	2.12 – 4.32	[110, 128]
		CS	Longitudinal	8.32 ± 6.44	[110]
	Humerus	CT	Longitudinal	1.77	[129]
Bovine	Femur	CT	Longitudinal	3.615 ± 0.728	[130]
		SENT	Longitudinal	3.21	[131]
		SENB	Longitudinal	2.30 – 6.2	[105] [132]
		CT	Radial	4.01 ± 0.31	[106]
		CST	Radial	3.29 ± 0.48	[106]
		CT	Transverse	2.4 – 5.2	[133]
		CNB	Transverse	5.8 ± 0.5	[134]
		SENB	Transverse	3.48 ± 0.33	[105]
		SEVN (no precrack)	Transverse	4.75 ± 0.975	[135]
		SEVNB	Transverse	5.5 ± 0.6	[136]
		SENT	Transverse	5.58	[131]
		SENB	Transverse	5.1 ± 0.5	[132]
		SENB	Transverse	10.5 ± 0.5	[132]
	Tibia	CT (speeds 1E-3 to 5 cm min^{-1})	Longitudinal	2.8 – 6.3	[120]
		CT	Longitudinal	6.73 – 6.29	[128]
		SENT	Transverse	11.2 ± 2.6	[123]
		SENT	Transverse	2.2 – 4.6	[137]
		SENB	Transverse	4.53 ± 0.98	[105]
Baboon	Femur	CST	Longitudinal	1.75	[138]
		SENB, 3PB	Transverse	5.8 – 7.3	[73]
		CST	Longitudinal	1.73 – 2.25	[72]
		SENB	Transverse	6.22	[138]
Canine	Femur	SENB	Transverse	6.79 ± 0.41	[70]
Equine	3rd Metacarpal	CT	Transverse	4.38 – 4.72	[107]
Manatee	Rib	CNB	Transverse	4.5 ± 0.5	[134]

Table 4.10: (b) Fracture toughness of cortical bone as measured by G_c.

Species	Bone	Test	Orientation	G_c (N m^{-1})	Reference
Human	Dentin	CT	Longitudinal	742 ± 137	[126]
	Femur – Lateral	CT	Longitudinal	72.5 – 74.5	[124]
	Femur – Medial	CT	Longitudinal	74.5 – 74.6	[124]
	Femur – Lateral	CS	Longitudinal	74.5 – 74.6	[124]
	Femur – Medial	CS	Longitudinal	66.0 – 80.0	[124]
	Femur	SENB	Transverse	G_{CI} = 200; G_{CII} = 50	[139]
	Tibia	CT	Longitudinal	595 – 827	[128]
	Tibia	CT	Longitudinal	339 ± 132	[110]
	Tibia	CS	Longitudinal	4200 ± 2516	[110]
	Tibia – Lateral	CT	Longitudinal	68.2 – 78.5	[124]
	Tibia – Medial	CT	Longitudinal	72.1 – 77.0	[124]
	Tibia – Lateral	CS	Longitudinal	67.7 – 78.5	[124]
	Tibia – Medial	CS	Longitudinal	72.1 – 78.5	[124]
	Femur	SENB	Longitudinal	G_{CI} = 1; G_{CII} = 12	[139]
Bovine	Tibia	CT	Longitudinal	987 – 896	[128]
		CT (speeds 10^{-3} to 5 cm min^{-1})	Longitudinal	630 – 2884 J m^{-2}	[120]

propagation [114]. Furthermore, the capacity of tissue to absorb impact energy has been shown to be negatively correlated to osteon density [115].

The higher value of fracture toughness in the transverse directions of bone is due to the following toughening mechanisms: (1) soft cement lines that can help deviate crack propagation towards itself, thereby influencing the fracture toughness of bone [116,117]; (2) debonding at cement lines provides an additional mechanism of energy dissipation for transverse cracks [118,119], and (3) bridging effect of the osteons at the crack plane, thus requiring additional energy to pullout the osteons during the crack propagation in the direction [120–123].

Porosity has also been shown to contribute to the fracture toughness of bone. This contribution is influenced by the direction of crack propagation and the loading mode. In the longitudinal direction, there is an inverse relationship between porosity and fracture toughness [124]. This may be due to the net decrease in crack face area. Porosity has little influence on transverse fracture toughness [72], which has been shown to be influenced predominately by osteon pullout and cement line debonding.

Table 4.10: (c) Toughness of cortical bone as measured by work to failure.

Species	Bone	Test	Orientation	W_F (kJ/m^2)	Ref.
Human	Femur	3pt-Bending	Transverse	12.7 ± 3.2	[55]
		Impact	Transverse	2 – 5	[140]
		SENB	Transverse	1.7 – 3.3	[127]
		3pt-Bending	Transverse	62.9 ± 19.5	[56]
Bovine	Femur	CT	Longitudinal	1.27 ± 0.56	[130]
		SENB	Longitudinal	0.40 ± 0.07	[105]
		SENB	Transverse	1.15 ± 0.19	[105]
		SEVNB	Transverse	7.1 ± 1.4	[136]
		CT	Transverse	0.92 – 2.78	[133]
	Tibia	SENB	Transverse	1.47 ± 0.49	[105]
Equine	3rd Metacarpal	4pt-Bending	Transverse	22.5 ± 0.77	[71]
		Impact	Transverse	13.0 ± 0.64	[71]
Baboon	Femur	SENB, 3pt-Bending	Transverse	115 – 181	[73]

Notes for Table 4.10: Longitudinal – crack runs parallel to osteons; Transverse – crack runs perpendicular to osteons.
CT: Compact Tension; 3PB: 3-Point Bending (no notch introduced); CS: Compact Shear; SENB: Single Edge Notched Bending; CNB: Chevron Notched Beam; SENT: Single Edge-Notched Tension; CST: Compact Sandwich Toughness; SEVN: Single Edge V-Notched; SEVNB: Single Edged V-Notched Beam.

The fracture toughness measurements of trabecular bone (Table 4.11) have been limited to a few studies [141]. It was reported that the critical stress intensity factor was determined using transverse and longitudinal compact tension and single edge-notched bending specimens from osteoporotic and osteoarthritic human femoral heads and from non-pathological horse vertebrae. Fracture toughness values exhibited wider variations for the human osteoporotic and osteoarthritic specimens compared to the normal specimens. Further, there was a positive correlation between fracture toughness and apparent density. If expressing the relationship in the form of a power function, the exponents range from 1.44 to 1.62. The wide variation in properties leads to the possibility that apparent density is not the sole contributor to fracture toughness.

4.3.4   VISCOELASTIC BEHAVIOR OF CORTICAL AND TRABECULAR MACROSCOPIC BONE

Cortical bone exhibits a rate-sensitive behavior to mechanical loading. An early study [60] of human femoral bone showed compressive stiffness and strength to increase by factors 2.7 and 2.1, respectively, as loading rate increased from 0.001 to 1500s^{-1}. Over the same loading rates the compressive stiffness and strength of bovine femoral bone increased by factors of 2.3 and 2.1, respectively. Poisson's ratio for human bone declined from 0.30 to 0.26 over loading rates of 0.001 to 300s^{-1}. A study of human tibial bone over compressive loading rates of 0.001 to 10s^{-1} also showed stiffness to increase by 150% and strength by 220% [88]. Similar increases in stiffness and strength were evident

Table 4.11: Fracture behavior of trabecular bone.

Species	Bone	Test	Fracture Direction	K_{max} (MPa m$^{0.5}$)	Ref.
Human Osteoporotic	Femoral Head	SENB	Perpendicular, Aligned	0.07 – 0.5, 0.04 – 0.3	[141]
		CT		0.08 – 0.6 0.08 – 0.5	
Human Osteoarthritic	Femoral Head	SENB		0.11 – 0.7, 0.09 – 0.5	
		CT		0.11 – 0.9 0.17 – 0.9	
Equine	Vertebrae	SENB		0.1 – 0.35, 0.09 – 0.5	
	Vertebrae	CT		0.42 – 0.9, 0.22 – 0.83	

in torsional loading of rabbit femurs which exhibited 33% more torque and torsional deformation and 5% higher stiffness when the highest loading rate was compared to the lowest [142].

It has been shown that the elastic modulus, strength, and strain at maximum stress increase, while yield stress/strain decrease with loading rate when human cortical bone is loaded in compression. In contrast, the elastic modulus of human cortical bone in tension increases but its strength, strain at maximum stress, and yield stress/strain all decrease with the loading rate. In addition, the post-yield deformation of human cortical bone is more sensitive to strain rate than to the initiation of macroscopic yielding [143]. Bone specimens loaded at higher strain rates usually behave in a more brittle manner, while those loaded at lower rates become tougher. One idea is that low strain rates allow more time for micro-cracks to develop, thereby increasing the toughness of bone [144]. Additionally, cortical bone's ability to delay the ductile to brittle transition is thought to be an important factor underlying the toughness of the tissue.

The rate-sensitivity of cortical bone mechanical properties does not exhibit a distinct influence by its structural anisotropy. This has been verified by experimental studies. For example, though transverse bovine tibial specimen stiffness was significantly lower than that of longitudinal specimens, both orientations exhibited relatively constant changes in values over loading rates from 0.001 to 1000s^{-1} [145]. Strength values were increased by factors of 1.78 and 2.0 in longitudinal and transverse orientations, respectively. In compression tests of bovine femurs (loading rates of 300 to 3.5×10^5 psi/s) [146], radial stiffness increased by a factor of 2.6, whereas those in the circumferential and longitudinal orientations increased by 1.6 and 1.2, respectively. Strength values did not exhibit the same sensitivity to anisotropy, with values for the three orientations increasing by factors 1.13, 1.26, and 1.35, respectively, (circumferential, radial, and longitudinal).

Dynamic mechanical analysis of bone has indicated that the hydration condition, but not the collagen phase, has a significant effect on loss factor, with wet samples exhibiting higher values than dry samples [147]. In addition, hydration state but not collagen denaturation strongly influenced the storage modulus of bone [148]. These findings indicate that moisture content may have a significant role in bone viscoelasticity.

The viscoelastic behavior of trabecular bone is influenced by several factors, which include the type of specimen tested (size and presence of marrow) as well as the loading methodology. It has been pointed out that the viscoelastic response of trabecular bone in compression may depend on both the presence of marrow within the tissue and the properties of the tissue itself [149]. Early studies on the hydraulic stiffening effect of bone marrow failed to find an effect [150, 151] or found one only at a very high loading rate [88]. However, another study [152] concluded that the presence of marrow does contribute to the strain-rate sensitivity of trabecular bone. The contradictory results between studies may be due to the testing of specimens which confined the marrow [150] versus those that did not [151].

Determining the viscoelastic properties of trabecular tissue itself have also produced mixed results. An early DMA study [151] of trabecular bone from the human femoral head subjected to small amplitude excitations over frequencies of 200 to 3,000 Hz failed to demonstrate an appreciable viscous response for either fresh wet bone or defatted bone. Similarly, no viscous effect on stiffness was noted for femoral specimens loaded at strain rates ranging from 10^{-4} to $10^2 s^{-1}$ [153]. However, compression tests of trabecular bone loaded at strain rates from 0.001 to $10 s^{-1}$ showed compressive stiffness and strength to increase by factors 1.5 and 2.2, respectively, as the loading rate increased [88]. Stiffness and modulus are proportional to the strain rate raised to the power of 0.06. Over the same range of strain rates, stiffness and strength are proportional to apparent density raised to the power of 3 and 2, respectively [88]. A separate study [154] also found a power law strain-rate dependency for stiffness, strength, and ultimate strain with the exponents of 0.047, 0.073, and 0.03, respectively.

4.3.5 FATIGUE BEHAVIOR OF CORTICAL AND TRABECULAR MACROSCOPIC BONE

The fatigue behavior of cortical bone is an important area of research as it has been implicated in pathologies, such as stress fractures [155, 156] and femoral neck fractures [157]. Failure of bone in these pathologies is most likely due to a reduction in mechanical properties that results from fatigue loading. For example, tensile fatigue tests of bovine bone loaded to 50% of their fatigue life have shown stiffness losses of up to 20% at failure as well as a 13% loss of tensile strength [158]. Decreases of 7% and 20% in stiffness and Poisson's ratio have been observed over the fatigue life of human femoral bone subjected to load magnitudes corresponding to $2000\mu\varepsilon$ [159]. Tension and compression fatigue tests have shown reductions in secant modulus [160]. Additionally, increased cyclic energy dissipation at loading levels above $2500\mu\varepsilon$ in tension and $4000\mu\varepsilon$ in compression implicate these levels as critical damage thresholds for these loading modes [160].

Early research [161] into the fatigue behavior of human trabecular bone focused on determining the stiffness behavior of specimens during low cycle compression tests (0.1 Hz, 30 cycles) to 50% of predicted ultimate strength. Stiffness of the material was shown to increase for the first 10 cycles and then decrease. Subsequent research [162] seeking to determine the fatigue life of bovine trabecular bone loaded in compression found that the number of cycles to failure was strongly correlated with the initial global maximum. The number of cycles to failure ranged from 20 cycles at 2.1% strain to 400,000 cycles 0.8% strain.

It has also been shown that bovine trabecular bone is characterized by a decreasing modulus and increasing creep behavior prior to failure, with cycles to failure characterized by a power law relationship with an exponent of -11.19 [163]. A subsequent study [164] noted the similarity of this exponent to that of human femoral cortical bone (-12.22, [165]). Motivated by this, compression fatigue tests were performed on human vertebral specimens. After accounting for differences in mean monotonic yield strains between the bovine and human specimens, it was found that a single S-N curve could account for the pooled data of both species. Due to the structural difference between human and bovine bone, the findings suggested that the dominant fatigue failure mechanism might initiate at the ultrastructural level.

4.4   MECHANICAL PROPERTIES OF BONE AT THE MICROSCOPIC LEVEL

Testing of individual osteons, regions of interstitial tissues, and individual trabeculae comprise most microscopic level testing (on the order of microns). Cortical bone specimens can be dissected or machined into uniform shapes, allowing for a wide range of tests to be performed. At the microscopic level, individual trabecular are irregular in shape. This may increase the difficulty of obtaining accurate information from tension and compression tests, requiring the production of uniform specimens or the utilization of a bending test.

4.4.1   ELASTIC PROPERTIES OF CORTICAL AND TRABECULAR MICROSCOPIC BONE

Due to the high heterogeneity of bone structure, its elastic behavior is significantly dependent of the size of test specimens. Mechanical tests of human cortical bone at the microscopic level have focused primarily on assessing the mechanical behavior of osteons and interstitial bone tissues. Micro-specimens taken from osteons and interstitial tissues of human femurs have revealed significantly different properties when compared to those of macro-specimens (Table 4.12). The tensile [166] and compressive [167] moduli at the microscopic level are much lower than those obtained at the macroscopic level. In contrast, torsion tests at microscopic level have yielded modulus values significantly higher than those obtained at the macroscopic level [168]. Further, within each testing mode, scattering of elastic modulus values is evident and attributable to the varied degree of mineralization of the osteons that have different biological ages and the alternating orientation of collagen fibrils in the osteons.

Mechanical tests of the elastic properties of individual trabecular specimens have been limited to 3pt-bending tests on specimens machined from human tibia [170,171]. It is observed that elastic properties of individual trabeculae are significantly different form those of macroscopic specimens.

Table 4.12: Micro-elastic properties of cortical bone.

Species	Bone	Test	Orientation	Modulus (GPa)	Comments	Ref.
Human	Femur	Tension	Longitudinal	3.9 – 11.7	Osteons	[166]
		Compression	Longitudinal	3.3 – 9.3	Osteons	[167]
		Torsion	Longitudinal	17.2 – 23.2	Osteons	[168]
		3pt-Bending	Longitudinal	2.3 – 2.7	Osteons	[169]
	Tibia	3-pt-Bending	Longitudinal	5.4	Micro-beams	[170]
		4pt-Bending	Longitudinal	6.8	Micro-beams	[171]
Ox	Femur	Tension	Longitudinal	5.4 – 14.7	Osteons	[166]
Mouse	Femur	4pt-Bending	Longitudinal	7.9 / 11.4	Posterior / Anterior Quadrants	[172]

Table 4.13: Elastic behavior of trabecular bone.

Species	Bone	Test	Specimen Orientation	Stiffness (MPa)	Ref.
Human	Tibia	3pt-Bending	SI	4870 ± 1840	[170]
		3pt-Bending	ML	3830 ± 450	[170]
		3pt-Bending	V & H	5720 ± 1.27	[171]

SI: Superior-Inferior; V: Vertical; H: Horizontal; L: Longitudinal; ML: Medio-Lateral

Vertically oriented specimens have exhibited higher stiffness values compared to their horizontal counterparts [170]. The stiffness values of trabeculae exist at the lower range of values reported for macroscopic specimens. Similarly-sized cortical specimens (which included several osteons) exhibited 18% higher stiffness compared to trabecular tissue [171].

4.4.2 POST-YIELD BEHAVIOR OF CORTICAL AND TRABECULAR MICROSCOPIC BONE

Post-yield properties are also dependent on test specimen size. The mechanical properties of femoral osteons in tension [166], compression [167], and torsion [168] exhibit significant differences compared to those obtained from macroscopic test specimens. The range of strengths for the osteons is below that of macro-specimens in both tension and compression. In contrast, osteons tested in torsion exhibited higher strengths than macro specimens did. These trends are similar to those observed in the elastic properties of human cortical bone between macro and micro-specimens.

Post-yield tests of individual trabecular specimens have been limited to vertically oriented specimens dissected from human thoracic vertebrae [173]. Ultimate strains average 8.8%. Significant intra-individual variations in ultimate-strains were observed, ranging from 1.8% to 20.2%. Significant variations were also found within donors, with standard deviations in ultimate-strain ranging from

Table 4.14: Post-yield micro properties of cortical bone.

Species	Bone	Test	Specimen Orientation	Strength (MPa)	Ultimate Strain %	Comm.	Ref.
Human	Femur	Tension	Longitudinal	84 – 120	6.8 – 10.5	Osteons; 20 & 80 y.o. Males	[166]
		Compression	Longitudinal	78.4 – 163.8	1.9 – 3.0	Osteons	[167]
		Torsion	Longitudinal	170.9 – 205.6	4.6 – 11.6 degs.	Osteons	[168]
Ox	Femur	Tension	Longitudinal	110.7 – 118.9	8.3	Micro. Specimen	[166]
Mouse	Femur	4pt-Bending	Longitudinal	187.1 / 196.7		Posterior / Anterior	[172]

15% to 58% of the mean values. This heterogeneity at the micro-scale may underlie the large variations in properties that are observed at the macro-scale. Failure of macroscopic specimens most likely initiates at the weakest trabeculae, which by their wide variation in post-yield properties would tend to drive the macroscopic behavior.

4.5 MECHANICAL PROPERTIES OF BONE AT THE SUB-MICROSCOPIC LEVEL

Mechanical properties of osteonal and interstitial lamellae have been assessed using nanoindentation techniques. These studies have shown tissue type and degree of mineralization to significantly affect stiffness and hardness properties. Tests of human femur and tibia specimens have shown osteonal tissue to exhibit lower stiffness and hardness than interstitial tissues [174, 175]. This is consistent with the mineralization process in that interstitial tissues are chronologically older than osteons and thereby are more likely to be fully mineralized. Testing of femoral osteonal tissue has shown an outwards decline and increase in elastic modulus from the center of the osteon to the cement line by Rho et al. [175] and Gourion-Arsiquaud et al., respectively. These contradictory results are not readily explained since there are also conflicting studies on the mineralization process of secondary osteons [176, 177]. A study in which osteons from bovine tibia were progressively demineralized showed stiffness to decrease as the weight percentage of mineral decreased [178]. This is consistent with macroscopic studies where the mineral phase has been shown to be the main contributor to bone stiffness [73].

Indentation orientation has also been shown to have a significant influence on stiffness and hardness values of cortical bone. Osteons have shown anisotropic behavior with regard to elastic properties [179–181]. Elastic modulus of osteons are significantly higher in the longitudinal than the transverse direction [179, 180]. Anisotropic behavior has also been observed for shear moduli determined by nanoindentation [181]. Additionally, interstitial and osteonal tissues that are normally

Table 4.15: Sub-microscopic properties of cortical and trabecular bone. *(Continues.)*

Species	Bone	Orientation	Modulus (GPa)	Hardness (GPa)	Comments	Ref.
			Cortical Bone			
Human	Femur	Longitudinal	20.8, 18.8	0.65, 0.55	Osteonal – Dry; Inner, outer lamellae	[175]
		Longitudinal	24.2 – 26.8	0.72 – 0.86	Interstitial - Dry	[175]
		Longitudinal	23.45 ± 0.21		Osteonal - Dry	[179]
		Transverse	16.58 ± 0.32		Osteonal - Dry	[179]
		Longitudinal	18 ± 1.7	0.6 ± 0.11	Osteonal - Dry	[185]
		Longitudinal	16.5 – 24		Osteonal – Dry	[183]
		Longitudinal	9.5 – 19.75		Osteonal – Wet	[183]
		Longitudinal	18.6 ± 4.2	0.52 ± 0.15	Osteons – Wet	[184]
		Longitudinal	20.3 ± 5.1	0.59 ± 0.20	Interstitial - Wet	[184]
		Radial, Longitudinal, Circumferential	9.17, 17.28, 24.66		Osteons – Dry	[181]
		Radial, Longitudinal, Circumferential	4.69, 5.61, 7.68		Osteons – Dry; Shear Moduli	[181]
	Tibia	Longitudinal	22.5 ± 1.3	0.614 ± 0.042	Osteons - Dry	[174]
		Longitudinal	25.8 ± 0.7	0.736 ± 0.034	Interstitial - Dry	[174]
		Longitudinal	19.0 ± 1.7, 17 + 1.5		Osteonal – 90% Compression, Tension – Dry	[182]
		Longitudinal	21.4 ± 0.8, 18.9 ± 1.0		Interstitial – 90% Compression, Tension – Dry	[182]
	Metatarsal	Longitudinal	20.1 ± 1.1, 17.3 ± 0.9		Interstitial, Osteonal 90% Compression – Dry	[182]
Bovine	Tibia	Transverse	12.9, 3.0, 1.9		Osteonal 58, 26, 0 Wt% Mineral Content - Dry	[178]
Horse	Radius	Longitudinal	16.5 – 23.75		Osteonal - Dry	[180]
		Transverse	10.2 – 15.7		Osteonal - Dry	[180]
			Trabecular Bone			
Human	Vertebrae	Transverse	13.4 ± 2.0	0.468 ± 0.079	Trabecula – Dry	[174]
	Femur	Unknown	18.14 ± 1.7		Trabecula - Dry	[179]
		Transverse	22.5 ± 3.1	1.1 ± 0.17	Trabecula - Dry	[185]
		Unknown	21 – 27.5		Trabecula – Dry	[183]

Table 4.15: *(Continued.)* Sub-microscopic properties of cortical and trabecular bone.

					Newborn (112 – 115 Post Conception) - Dry	
Rabbit	Femur	Oblique	20	0.83 ± 0.06	Cancellous Lamellar – Dry	[187]
		Oblique	16	0.82 ± 0.09	Cancellous Interlamellar - Dry	[187]

under physiologically compressive loads have been shown to be stiffer than those that are under tensile loads [182].

The state of hydration of the cortical bone can significantly influence stiffness and hardness. Many of the studies reviewed were conducted under dry conditions. Studies in which bone was tested under dry and wet conditions have shown dry osteonal tissue to have a significantly higher stiffness than wet bone [183]. Further, it has been shown that when wet, interstitial tissue continues to be stiffer and harder compared to osteonal tissue [184].

Stiffness and hardness properties have also been determined for single trabeculae using nanoindentation techniques. Dry tests of individual trabeculae have shown wide variations in the stiffness [174,179,185]. These stiffness values are comparable to values that have been observed for cortical tissue. Anatomic location also influenced stiffness values with trabeculae from the femur [179] exhibiting higher values compared to those from the vertebrae. As with cortical bone, dry trabecular bone has shown to be stiffer and harder than wet bone [183]. Similar to cortical bone, trabecular bone demonstrates increases in stiffness with increasing mineralization [186].

4.6   MECHANICAL PROPERTIES OF HYDROXYAPATITE AND COLLAGEN

Tests have been conducted to measure the mechanical properties of synthetic and naturally occurring single crystal hydroxyapatite by nanoindentation [188, 189]. Synthetic crystals exhibited stiffness values of 125.9 and 135.1 GPa and hardness values of 8.8 and 9.7 GPa for the side and basal planes, respectively [188]. Observed fracture toughness values for the side and basal plane were of 0.65 and 0.40 MPa m$^{1/2}$, respectively [188]. Large, naturally occurring single millimeter-sized crystals have shown higher stiffness (150.4 GPa versus 143.6 GPa) and hardness (7.1 versus 6.4 GPa) for the base versus side, respectively [189]. However, cracking initiated earlier on the base compared to the side with fracture toughness values of 0.45 and 0.35 MPa m$^{1/2}$, respectively [189]. Thus, the basal plane is harder and stiffer than the side, but not as tough. These difference are evidence of the anisotropic nature of hydroxyapatite and could have implications for the anisotropy seen at higher length scales. Additionally, differences in fracture toughness observed between the basal and side planes could influence crack propagation of cracks in bone.

Utilizing a micro-electro-mechanical system (MEMS), mechanical tests have been conducted on individual Type I collagen fibrils obtained from sea cucumber dermis [190]. These collagen fibrils were ~12μm long with diameters of 150~470nm. They also assemble with the same repeat period, gap to overlap ratio, and crosslinking chemistry as vertebrate collagen. Tensile loading of partially hydrated fibrils showed the average elastic modulus to be 0.86 GPa for strains less than 0.09. Yield strength and yield strain were 0.22 GPa and 0.21, respectively. The structure of the fibril was shown to affect mechanical properties even at this small scale. It was noted that as the diameter of the fibril increased, the yield strength decreased. It was speculated that molecular defects and inhomogeneity in cross-linking density might have been responsible for the observed behavior.

Atomic force microscopy has also been utilized to study the interaction between collagen fibrils and intrafibrillar mineral content [191]. For both fully and partially demineralized fibrils, the modulus measurements varied along the fibril axis in a periodic manner coincident with the gap regions and overlap zones of the fibril's structure. For recombinant collagen (which was never mineralized to begin with), the modulus values were approximately 30 MPa at the gaps and 60 MPa near the center of the overlaps. Fully demineralized human dentin collagen fibrils exhibited lower modulus values of 10 MPa at the gaps and 40 MPa at the overlaps. Dentin collagen fibrils subjected to 240 sec. of demineralization (which partially demineralized the fibril) had gap modulus values of 200~400 MPa and overlap values 500~800 MPa. Smooth dentin in which the fibrils were essentially fully mineralized exhibited gap and overlap moduli of 1.2 GPa and 1.5 GPa, respectively. The periodicity of modulus as well as increasing stiffness values with mineralization underscore the role of structure and degree of mineralization in the mechanical properties of collagen.

Hydration state has also been shown to have a significant influence on the mechanical properties of collagen fibrils. Fully hydrated and demineralized dentin from human teeth has been shown to exhibit viscoelastic behavior with relaxed elastic modulus of approximately 148.7 KPa, relaxation time constant for constant strain 5.1 seconds, and relaxation time constant for constant stress of 6.6 seconds [192]. When desiccated, demineralized dentin loses its viscoelastic properties and exhibits a stiffness of 2.1 GPa and hardness of 0.2 GPa [192]. Upon rehydration, the demineralized dentin recovers some, but not all, of its original viscoelasticity. Relaxed elastic modulus, constant strain time constant, and constant stress time constant recovered to 381.4 KPa, 2.9 seconds, and 3.2 seconds [192]. The inability of the modulus to fully recover upon rehydration was attributed to chain entanglement and oxidation-induced crosslinking [192].

4.7   PHYSIOLOGICAL RESPONSE OF BONE TO FUNCTIONAL LOADING

The design of tissue engineered constructs whose purpose is to repair defects in load-bearing hard tissues requires knowledge of the loading environment the tissue normally experiences. Numerous studies have been conducted (Table 4.16) which assess the *in vivo*, organ level stress and strain response of bone to physiological loading. *In vivo* studies have been conducted on humans. However,

Table 4.16: In vivo strain measurements in humans and animals (Reprodued from [13]).
(Continues.)

Species	Bone	Aspect	Activity	Principal or Max. Strain ($\mu\varepsilon$)	Max Strain Rate ($\mu\varepsilon/s$)	Ref.
Human	Femur	Proxima-lateral	Two-legged stance	494T, 347C		[197]
			One-legged stance	1,463T, 435C		[197]
			Walking	1,198T, 393C		[197]
			Stair Climbing	1,454T, 948C		[197]
	Tibia	Antero-medial	Walking	400 C	4000C	[193]
			Running	850 T	13000T	[193]
		Medial	Walking	540C, 440T, 870S	7,200C, 11,000T, 16,200S	[195]
			Walking with Pack	560C, 380T, 770S	6,400C, 11,400T, 15,500S	[195]
			Jogging	880C, 630T, 1440S	27,400C, 13,900T, 38,900S	[195]
			Sprinting	970C, 650T, 1580S	34,500C, 20,200T, 51,400S	[195]
Horse	Radius	Lateral-Caudal, Cranial- Medial	Standing	1500C 900T		[198]
			Walking	1900C 1200T		[198]
			Pacing	2600C 1500T		[198]
	Tibia	Lateral- Caudal, Cranial- Medial	Standing	300C 80T		[198]
			Walking	950C 820T		[198]
			Trotting	1600C 1500T		[198]
		Caudal	Slow Walk	940C		[194]
			Fast Walk	1300C		[194]
			Trot	1940C		[194]
			Fast Canter	3150C		[194]
	Radius	Caudal	Slow Walk	1780C		[194]
			Fast Walk	1970C		[194]
			Trot	2630C		[194]
			Fast Canter	2320C		[194]

Table 4.16: *(Continued.)* In vivo strain measurements in humans and animals (Reproduced from [13]). *(Continues.)*

Sheep	Radius	Cranial	Walking (1 m/s)	640~930T, 360C	6,700~14,700	[202, 205]
		Caudal	Walking (1 m/s)	480T 1,170~1,320C	10,000~17,900	[202, 205]
	Tibia	Medial	Walking	262		[199]
			Trotting	305		[199]
		Cranial	Walking (1 m/s)	709T, 453C	17,800	[206]
		Caudal	Walking (1 m/s)	378T, 666C	16,200	[206]
	Calcaneus	Dorsal	Slow Walk	80T, 170C, 166.6S	2,700	[200]
			Medium Walk	100T, 210C, 200S	3,600	[200]
			Slow Trot	100T, 220C, 170S	4,300	[200]
			Medium Trot	120T, 250C, 230S	5,100	[200]
Goat	Radius	Caudal	Slow Walk	820C		[201]
			Fast Walk	830C		[201]
			Fast Trot	1750C		[201]
			Max Gallop	1850C		[201]
	Tibia	Caudal	Slow Walk	725C		[201]
			Fast Walk	850C		[201]
			Fast Trot	1800C		[201]
			Max Gallop	1970C		[201]
Pig	Radii	Craniomedial	Walking	837C	30000	[207]
Dog	Femur	Medial,lateral	Walking	240T, 460C		[203]
		Medial	Walking	246.6T, 413.2C		[204]
	Radius	Caudal	Fast Walk	1500C		[194]
			Fast Trot	2380C		[194]
			Fast Canter	2250C		[194]
	Tibia	Caudal	Fast Walk	1060C		[194]
			Fast Trot	2000C		[194]
			Gallop	2020C		[194]
Macaque	Ulna	Lateral	Walking	750T, 360C		[208]
		Posterior	Walking	400T, 430C		[208]
		Medial	Walking	450T, 870C		[208]

Table 4.16: *(Continued.)* In vivo strain measurements in humans and animals (Reproduced from [13]).

Gibbon	Ulna	Ventral	Brachiating	1420T		[209]
		Dorsal	Brachiating	1270T		[209]
	Radius	Ventral	Brachiating	1640C		[209]
		Dorsal	Brachiating	750T		[209]
	Humerus	Ventral	Brachiating	420T		[209]
		Dorsal	Brachiating	1500T		[209]
Rat	Femur	Anerior-Lateral	Exercise Wheel	410C, 250T	20000	[210]
		Anerior-Lateral (Transverse)	Exercise Wheel	260C, 120T, 420S		[210]
	Ulna	Medial	Running	1200C	23000, 38000	- [211]
			Height Dropping	2500C	-100000	[211]
Potoroos	Calcaneus	Lateral	Walking	2000C		[212]
Turkey	Ulna	Midshaft	Wing flapping	3300C	56000	[196]
Rooster	Tarsometatarsal	Anterior(C), Posterior (T), Anterior (S)	Running	1850C, 1220T, 1580S		[213]

Note: "C" represents compression, "T" tensions, and "S" shear.

the manner in which this data is obtained has necessitated a reliance on the use of animal studies to provide the majority of this information.

It has been pointed out that there are similar ranges of strain experienced by the bone under physiological loading conditions across multiple species [13]. For example, humans have been shown to exhibit strains similar to other animals [193], with peak functional strains falling between -2000 and -3200$\mu\varepsilon$ [194]. Peak strain rates have been shown to be up to -100,000$\mu\varepsilon$/s for rats dropped from a height and 64,100$\mu\varepsilon$/s for galloping horses [194]. However, these conditions are atypical. For more common circumstances, peak strain rates have ranged from -34,500$\mu\varepsilon$/s [195] to 56,000$\mu\varepsilon$/s [196]

A review of the physiological loading literature reveals that, as expected, the maximal strains and loading rates experienced by bone depend on the activity. For example, the maximal strains experienced by the femur in single-legged stance is significantly higher than observed in a two-legged stance [197]. Vigorous activities such as stair climbing produce higher strains when compared to less vigorous activities such as walking [197]. Tibial strains can increase two-fold and strain rates nearly five-fold as the activity transitions from walking to sprinting [195]. Similar observations are evident in other species such as the horse, [194,198], sheep [199,200], goat [201], and dog [194].

Another observation that can be made is that for a given activity the amount of strain experienced by a bone can vary significantly depending on the anatomic site. In the horse, strains in the radius are generally higher than those in the tibia [194,198]. However, there is activity dependence in that the differences between the two sites decrease as the activity becomes more vigorous (standing to trotting). Similar behavior is also seen between the radius and tibia of sheep [199,202], goats [201], and amongst the femur [203,204], radius, and tibia [194] of dogs.

A final issue to consider is that within the same bone, significant differences may exist depending on where along the shaft (in long bones) and in what quadrant (anterior, posterior, medial, or lateral) the measurements are taken. For example, in the walking human the antero-medial aspect of the tibia experiences lower compressive strains than the medial [193,195]. In sheep, the cranial portion of the tibia experience high strains during walking than the caudal.

4.8   CONCLUDING REMARKS

Due to the complex hierarchy of bone tissues, the amount of information in the literature on each topic covered in this chapter is vast and the provided tables and associated commentary by no means exhaustive. The preceding chapter illustrates that the mechanical behavior of bone is affected by multiple factors. Most important among them are hierarchical level, loading mode, anatomic location, specimen orientation, and species.

The hierarchical structure of bone means that mechanical tests above the level of bone's individual material components are not truly material tests and involve structural features that may influence the observed mechanical properties. Regarding the mechanical properties of whole bones, mechanical tests at this level are tests of the bone as a structure and not as a material per se. This makes quantitative comparisons between studies difficult. However, useful qualitative relationships can be made which can provide information regarding the relationship of mechanical properties to bone structure and material composition.

At the tissue level, bone can be classified as cortical bone or trabecular bone. Each of these tissue types exhibits significant variations in properties when comparisons are made between hierarchical levels, anatomic locations, specimen orientations, and species. Differences in hierarchical level often involve the presence of structural features at one level that are not present at another level. Furthermore, as testing proceeds towards smaller specimens, consideration must be given to size effects and the accuracy and resolution of instrumentation at smaller scale. Differences in anatomic location may be a response of bone to the physiological loading it experiences at a given location. This is manifested in part by orientation effects that were noted for both cortical and trabecular bone. Differences in orientation for cortical bone are in large measure a consequence of its structural transverse isotropy as well as remodeling. For trabecular bone, differences may be due in part to the adaptive trabecular remodeling that seeks to organize structure and material distribution at locations that best resist physiological loading. Differences in species can be explained in large measure by differences in bone structure and material composition. However, similarities in properties between species that possess bone with very different structures may point to the influence of ultrastructural factors as the underlying basis for the observed similarities.

The response of human and animal bone to physiologically relevant loading is important for tissue engineers to understand in order to design and fabricate bone tissue engineering products. For example, many tissue engineering constructs are used in load-bearing applications. This information bridges the gap between real world deformations experienced by bone and the loading environment for which a particular implant or scaffold must be designed.

REFERENCES

[1] Rho, J-Y, Kuhn-Spearing L, Zioupos P. Mechanical properties and the hierarchical structure of bone. Medical Engineering & Physics 1998;20: 92–102. DOI: 10.1016/S1350-4533(98)00007-1 4.1

[2] Weiner, S, Wagner HD. THE MATERIAL BONE: Structure-Mechanical Function Relations. doi:10.1146/annurev.matsci.28.1.271. Annual Review of Materials Science 1998;28: 271–298. DOI: 10.1146/annurev.matsci.28.1.271 4.1, 4.2

[3] Chavassieux, P, Seeman E, Delmas PD. Insights into material and structural basis of bone fragility from diseases associated with fractures: how determinants of the biomechanical properties of bone are compromised by disease. Endocr Rev 2007;28: 151–64. DOI: 10.1210/er.2006-0029 4.2

[4] Jee, WS, Ma Y. Animal models of immobilization osteopenia. Morphologie 1999;83: 25–34. 4.2

[5] Mosekilde, L. Assessing bone quality–animal models in preclinical osteoporosis research. Bone 1995;17: 343S-352S. DOI: 10.1016/8756-3282(95)00312-2 4.2

[6] O'Loughlin, PF, Morr S, Bogunovic L, Kim AD, Park B, Lane JM. Selection and development of preclinical models in fracture-healing research. J Bone Joint Surg Am 2008;90 Suppl 1: 79–84. DOI: 10.2106/JBJS.G.01585 4.2

[7] Henderson, CN, Cramer GD, Zhang Q, DeVocht JW, Fournier JT. Introducing the external link model for studying spine fixation and misalignment: part 2, Biomechanical features. J Manipulative Physiol Ther 2007;30: 279–94. DOI: 10.1016/j.jmpt.2007.03.002 4.2

[8] Henderson, CN, Cramer GD, Zhang Q, DeVocht JW, Fournier JT. Introducing the external link model for studying spine fixation and misalignment: part 1–need, rationale, and applications. J Manipulative Physiol Ther 2007;30: 239–45. DOI: 10.1016/j.jmpt.2007.01.006 4.2

[9] Nunamaker, DM. Experimental models of fracture repair. Clin Orthop Relat Res 1998: S56–65. 4.2

[10] Misof, K. Collagen from the osteogenesis imperfecta mouse model (oim) shows reduced resistance against tensile stress. Journal of Clinical Investigation 1997;100: 40–5. DOI: 10.1172/JCI119519 4.2

[11] Grabner, B. Age- and genotype-dependence of bone material properties in the osteogenesis imperfecta murine model (oim). Bone 2001;29: 453–7. DOI: 10.1016/S8756-3282(01)00594-4 4.2

[12] Jepsen, KJ, Goldstein SA, Kuhn JL, Schaffler MB, Bonadio J. Type-I collagen mutation compromises the post-yield behavior of Mov13 long bone. J Orthop Res 1996;14: 493–9. DOI: 10.1002/jor.1100140320 4.2

[13] Liebschner, MA. Biomechanical considerations of animal models used in tissue engineering of bone. Biomaterials 2004;25: 1697–714. DOI: 10.1016/S0142-9612(03)00515-5 4.2, 4.16, 4.7

[14] Currey, JD, Pitchford JW, Baxter PD. Variability of the mechanical properties of bone, and its evolutionary consequences. J R Soc Interface 2007;4: 127–35. DOI: 10.1098/rsif.2006.0166 4.2

[15] Currey, JD. Differences in the tensile strength of bone of different histological types. J Anat 1959;93: 87–95. 4.2

[16] Currey, JD. Mechanical properties of bone tissues with greatly differing functions. J Biomech 1979;12: 313–9. 4.2

[17] Cristofolini, L, Viceconti M, Cappello A, Toni A. Mechanical validation of whole bone composite femur models. Journal of Biomechanics 1996;29: 525–535. DOI: 10.1016/0021-9290(95)00084-4

[18] Martens, M, Van Audekercke R, De Meester P, Mulier JC. The geometrical properties of human femur and tibia and their importance for the mechanical behaviour of these bone structures. Arch Orthop Trauma Surg 1981;98: 113–20. DOI: 10.1007/BF00460798

[19] Boehm, HF, Horng A, Notohamiprodjo M, Eckstein F, Burklein D, Panteleon A, Lutz J, Reiser M. Prediction of the fracture load of whole proximal femur specimens by topological analysis of the mineral distribution in DXA-scan images. Bone 2008;43: 826–31. DOI: 10.1016/j.bone.2008.07.244

[20] Cristofolini, L, Viceconti M. Mechanical validation of whole bone composite tibia models. Journal of Biomechanics 2000;33: 279–288. DOI: 10.1016/S0021-9290(99)00186-4

[21] Lin, J, Inoue N, Valdevit A, Hang YS, Hou SM, Chao EY. Biomechanical comparison of antegrade and retrograde nailing of humeral shaft fracture. Clin Orthop Relat Res 1998: 203–13.

[22] Mosekilde, L, Mosekilde L. Normal vertebral body size and compressive strength: relations to age and to vertebral and iliac trabecular bone compressive strength. Bone 1986;7: 207–12.

[23] Mosekilde, L, Bentzen SM, Ortoft G, Jorgensen J. The predictive value of quantitative computed tomography for vertebral body compressive strength and ash density. Bone 1989;10: 465–70.

[24] Mosekilde, L, Mosekilde L. Sex differences in age-related changes in vertebral body size, density and biomechanical competence in normal individuals. Bone 1990;11: 67–73.

[25] McGee-Lawrence, ME, Wojda SJ, Barlow LN, Drummer TD, Bunnell K, Auger J, Black HL, Donahue SW. Six months of disuse during hibernation does not increase intracortical porosity or decrease cortical bone geometry, strength, or mineralization in black bear (Ursus americanus) femurs. Journal of Biomechanics 2009;42: 1378–1383. DOI: 10.1016/j.jbiomech.2008.11.039

[26] Kasra, M, Grynpas MD. Effect of long-term ovariectomy on bone mechanical properties in young female cynomolgus monkeys. Bone 1994;15: 557–61.

[27] Ayers, RA, Miller MR, Simske SJ, Norrdin RW. Correlation of flexural structural properties with bone physical properties: a four species survey. Biomed Sci Instrum 1996;32: 251–60.

[28] Akkas, N, Yeni YN, Turan B, Delilbasi E, Gunel U. Effect of medication on biomechanical properties of rabbit bones: heparin induced osteoporosis. Clin Rheumatol 1997;16: 585–95. DOI: 10.1007/BF02247799

[29] Grardel, B, Sutter B, Flautre B, Viguier E, Lavaste F, Hardouin P. Effects of glucocorticoids on skeletal growth in rabbits evaluated by dual-photon absorptiometry, microscopic connectivity and vertebral compressive strength. Osteoporos Int 1994;4: 204–10. DOI: 10.1007/BF01623240

[30] Hoshaw, SJ, Cody DD, Saad AM, Fyhrie DP. Decrease in canine proximal femoral ultimate strength and stiffness due to fatigue damage. J Biomech 1997;30: 323–9. DOI: 10.1016/S0021-9290(96)00159-5

[31] Fischer, KJ, Vikoren TH, Ney S, Kovach C, Hasselman C, Agrawal M, Rubash H, Shanbhag AS. Mechanical evaluation of bone samples following alendronate therapy in healthy male dogs. J Biomed Mater Res B Appl Biomater 2006;76: 143–8. DOI: 10.1002/jbm.b.30341

[32] Norrdin, RW, Simske SJ, Gaarde S, Schwardt JD, Thrall MA. Bone changes in mucopolysaccharidosis VI in cats and the effects of bone marrow transplantation: Mechanical testing of long bones. Bone 1995;17: 485–489.

[33] Biewener, AA. Bone strength in small mammals and bipedal birds: do safety factors change with body size? J Exp Biol 1982;98: 289–301.

[34] Cubo, J, Casinos A. Mechanical properties and chemical composition of avian long bones. Eur J Morphol 2000;38: 112–21.

[35] Ejersted, C, Andreassen TT, Oxlund H, Jorgensen PH, Bak B, Haggblad J, Torring O, Nilsson MH. Human parathyroid hormone (1–34) and (1-84) increase the mechanical strength and thickness of cortical bone in rats. J Bone Miner Res 1993;8: 1097–101.

[36] Jorgensen, PH, Bak B, Andreassen TT. Mechanical properties and biochemical composition of rat cortical femur and tibia after long-term treatment with biosynthetic human growth hormone. Bone 1991;12: 353–9.

[37] Feldman, S, Cointry GR, Leite Duarte ME, Sarrió L, Ferretti JL, Capozza RF. Effects of hypophysectomy and recombinant human growth hormone on material and geometric properties and the pre- and post-yield behavior of femurs in young rats. Bone 2004;34: 203–215. DOI: 10.1016/j.bone.2003.09.006

[38] Vanderschueren, D, Van Herck E, Schot P, Rush E, Einhorn T, Geusens P, Bouillon R. The aged male rat as a model for human osteoporosis: evaluation by nondestructive measurements and biomechanical testing. Calcif Tissue Int 1993;53: 342–7. DOI: 10.1007/BF01351841

[39] Simske, SJ, Guerra KM, Greenberg AR, Luttges MW. The physical and mechanical effects of suspension-induced osteopenia on mouse long bones. J Biomech 1992;25: 489–99.

[40] Jämsä, T, Jalovaara P, Peng Z, Väänänen HK, Tuukkanen J. Comparison of three-point bending test and peripheral quantitative computed tomography analysis in the evaluation of the strength of mouse femur and tibia. Bone 1998;23: 155–161.

[41] Wallace, JM, Rajachar RM, Allen MR, Bloomfield SA, Robey PG, Young MF, Kohn DH. Exercise-induced changes in the cortical bone of growing mice are bone- and gender-specific. Bone 2007;40: 1120–1127. DOI: 10.1016/j.bone.2006.12.002

[42] Brodt, MD, Ellis CB, Silva MJ. Growing C57Bl/6 mice increase whole bone mechanical properties by increasing geometric and material properties. J Bone Miner Res 1999;14: 2159–66.

[43] Gordon, KR, Perl M, Levy C. Structural alterations and breaking strength of mouse femora exposed to three activity regimens. Bone 1989;10: 303–12.

[44] Simske, SJ, Greenberg AR, Luttges MW. Effects of suspension-induced osteopenia on the mechanical behaviour of mouse long bones. J Mater Sci Mater Med 1991;2: 43–50. DOI: 10.1007/BF00701686

[45] Silva, MJ, Brodt MD, Uthgenannt BA. Morphological and mechanical properties of caudal vertebrae in the SAMP6 mouse model of senile osteoporosis. Bone 2004;35: 425–431. DOI: 10.1016/j.bone.2004.03.027

[46] Harrigan, TP, Jasty M, Mann RW, Harris WH. Limitations of the continuum assumption in cancellous bone. J Biomech 1988;21: 269–75. DOI: 10.1016/0021-9290(88)90257-6 4.3

[47] Zysset, PK, Goulet RW, Hollister SJ. A global relationship between trabecular bone morphology and homogenized elastic properties. J Biomech Eng 1998;120: 640–6. DOI: 10.1115/1.2834756 4.3, 4.3.1

[48] Petrtyl, M, Hert J, Fiala P. Spatial organization of the haversian bone in man. J Biomech 1996;29: 161–9. DOI: 10.1016/0021-9290(94)00035-2 4.3.1

[49] Hert, J, Fiala P, Petrtyl M. Osteon orientation of the diaphysis of the long bones in man. Bone 1994;15: 269–77. 4.3.1

[50] Jackson, SA. The fibrous structure of bone determined by x-ray diffraction. J Biomed Eng 1979;1: 121–2. 4.3.1

[51] Ko, R. The tension test upon the compact substance of the long bones of human extremities. J Kyoto Pref Med Univ 1953;53: 503–525. 4.3.1, 4.3.2

[52] Burstein, AH, Reilly DT, Martens M. Aging of bone tissue: mechanical properties. J Bone Joint Surg Am 1976;58: 82–6. 4.3.1, 4.3.2

[53] Sedlin, ED. A rheologic model for cortical bone. A study of the physical properties of human femoral samples. Acta Orthop Scand Suppl 1965: Suppl 83:1–77. 4.3.1

[54] Sedlin, ED, Hirsch C. Factors affecting the determination of the physical properties of femoral cortical bone. Acta Orthop Scand 1966;37: 29–48. DOI: 10.3109/17453676608989401 4.3.1

[55] Wang, X, Bank RA, TeKoppele JM, Mauli Agrawal C. The role of collagen in determining bone mechanical properties. Journal of Orthopaedic Research 2001;19: 1021–1026. DOI: 10.1016/S0736-0266(01)00047-X 4.3.1

[56] Nyman, JS, Roy A, Shen X, Acuna RL, Tyler JH, Wang X. The influence of water removal on the strength and toughness of cortical bone. J Biomech 2006;39: 931–8. DOI: 10.1016/j.jbiomech.2005.01.012 4.3.1

[57] Reilly, DT, Burstein AH. The elastic and ultimate properties of compact bone tissue. J Biomech 1975;8: 393–405. 4.3.1, 4.3.2

[58] Dempster, WT, Liddicoat RT. Compact bone as a non-isotropic material. Am J Anat 1952;91: 331–62. DOI: 10.1002/aja.1000910302 4.3.1

[59] Vincentelli, R, Grigorov M. The effect of Haversian remodeling on the tensile properties of human cortical bone. J Biomech 1985;18: 201–7. 4.3.2

[60] McElhaney, JH. Dynamic response of bone and muscle tissue. J Appl Physiol 1966;21: 1231–6. 4.3.2, 4.3.4

[61] Reilly, DT, Burstein AH, Frankel VH. The elastic modulus for bone. J Biomech 1974;7: 271–5.

[62] Jepsen, KJ, Davy DT. Comparison of damage accumulation measures in human cortical bone. J Biomech 1997;30: 891–4. DOI: 10.1016/S0021-9290(97)00036-5 4.3.2

[63] Keller, TS, Mao Z, Spengler DM. Young's modulus, bending strength, and tissue physical properties of human compact bone. J Orthop Res 1990;8: 592–603. DOI: 10.1002/jor.1100080416

[64] Smith, JW, Walmsley R. Factors affecting the elasticity of bone. J Anat 1959;93: 503–23.

[65] Simkin, A, Robin G. The mechanical testing of bone in bending. J Biomech 1973;6: 31–9.

[66] Burstein, AH. Contribution of collagen and mineral to the elastic-plastic properties of bone. The Journal of Bone and Joint Surgery 1975;57: 956–61.

[67] Martin, RB. The effects of collagen fiber orientation, porosity, density, and mineralization on bovine cortical bone bending properties. Journal of Biomechanics 1993;26: 1047–54. DOI: 10.1016/S0021-9290(05)80004-1

[68] Schryver, HF. Bending properties of cortical bone of the horse. Am J Vet Res 1978;39: 25–8.

[69] An, YH, Kang Q, Friedman RJ. Mechanical symmetry of rabbit bones studied by bending and indentation testing. Am J Vet Res 1996;57: 1786–9.

[70] Wang, X, Shanbhag AS, Rubash HE, Agrawal CM. Short-term effects of bisphosphonates on the biomechanical properties of canine bone. J Biomed Mater Res 1999;44: 456–60. DOI: 10.1002/(SICI)1097-4636(19990315)44:4%3C456::AID-JBM12%3E3.0.CO;2-9

[71] Reilly, GC, Currey JD, Goodship AE. Exercise of young thoroughbred horses increases impact strength of the third metacarpal bone. J Orthop Res 1997;15: 862–8. DOI: 10.1002/jor.1100150611

[72] Wang, XD, Masilamani NS, Mabrey JD, Alder ME, Agrawal CM. Changes in the fracture toughness of bone may not be reflected in its mineral density, porosity, and tensile properties. Bone 1998;23: 67–72. DOI: 10.1016/S8756-3282(98)00071-4 4.3.3

[73] Wang, X, Bank RA, TeKoppele JM, Hubbard GB, Athanasiou KA, Agrawal CM. Effect of collagen denaturation on the toughness of bone. Clin Orthop Relat Res 2000: 228–39. 4.5

[74] Cezayirlioglu, H, Bahniuk E, Davy DT, Heiple KG. Anisotropic yield behavior of bone under combined axial force and torque. J Biomech 1985;18: 61–9. DOI: 10.1016/0021-9290(85)90045-4

[75] Kopperdahl, DL, Keaveny TM. Yield strain behavior of trabecular bone. J Biomech 1998;31: 601–8. DOI: 10.1016/S0021-9290(98)00057-8 4.3.1, 4.3.2

[76] Martens, M, Van Audekercke R, Delport P, De Meester P, Mulier JC. The mechanical characteristics of cancellous bone at the upper femoral region. J Biomech 1983;16: 971–83. 4.3.1, 4.3.2

[77] Vahey, JW, Lewis JL, Vanderby R, Jr. Elastic moduli, yield stress, and ultimate stress of cancellous bone in the canine proximal femur. J Biomech 1987;20: 29–33. 4.3.1, 4.3.2

[78] Kuhn, JL, Goldstein SA, Ciarelli MJ, Matthews LS. The limitations of canine trabecular bone as a model for human: a biomechanical study. J Biomech 1989;22: 95–107. 4.3.1, 4.3.2

[79] Kang, Q, An YH, Friedman RF. Mechanical properties and bone densities of canine trabecular bone. J Mater Sci Mater Med 1998;9: 263–7. DOI: 10.1023/A:1008852610820 4.3.1, 4.3.2

[80] Cowin, SC. Wolff's law of trabecular architecture at remodeling equilibrium. J Biomech Eng 1986;108: 83–8. DOI: 10.1115/1.3138584 4.3.1

[81] Odgaard, A, Kabel J, van Rietbergen B, Dalstra M, Huiskes R. Fabric and elastic principal directions of cancellous bone are closely related. J Biomech 1997;30: 487–95. DOI: 10.1016/S0021-9290(96)00177-7 4.3.1

[82] Mosekilde, L, Mosekilde L, Danielsen CC. Biomechanical competence of vertebral trabecular bone in relation to ash density and age in normal individuals. Bone 1987;8: 79–85. DOI: 10.1016/8756-3282(87)90074-3 4.3.1, 4.3.2

[83] Turner, CH, Cowin SC, Rho JY, Ashman RB, Rice JC. The fabric dependence of the orthotropic elastic constants of cancellous bone. Journal of Biomechanics 1990;23: 549–561. 4.3.1

[84] Rice, JC, Cowin SC, Bowman JA. On the dependence of the elasticity and strength of cancellous bone on apparent density. J Biomech 1988;21: 155–68. 4.3.1

[85] Hayes, WC, Piazza SJ, Zysset PK. Biomechanics of fracture risk prediction of the hip and spine by quantitative computed tomography. Radiol Clin North Am 1991;29: 1–18. 4.3.1, 4.3.2

[86] Keller, TS. Predicting the compressive mechanical behavior of bone. J Biomech 1994;27: 1159–68. 4.3.2

[87] Rohlmann, A, Zilch H, Bergmann G, Kolbel R. Material properties of femoral cancellous bone in axial loading. Part I: Time independent properties. Arch Orthop Trauma Surg 1980;97: 95–102. DOI: 10.1007/BF00380706

[88] Carter, DR, Hayes WC. The compressive behavior of bone as a two-phase porous structure. J Bone Joint Surg Am 1977;59: 954–62. 4.3.4

[89] Kaneps, AJ, Stover SM, Lane NE. Changes in canine cortical and cancellous bone mechanical properties following immobilization and remobilization with exercise. Bone 1997;21: 419–23. DOI: 10.1016/S8756-3282(97)00167-1

[90] Acito, AJ, Kasra M, Lee JM, Grynpas MD. Effects of intermittent administration of pamidronate on the mechanical properties of canine cortical and trabecular bone. J Orthop Res 1994;12: 742–6. DOI: 10.1002/jor.1100120518

[91] Poumarat, G, Squire P. Comparison of mechanical properties of human, bovine bone and a new processed bone xenograft. Biomaterials 1993;14: 337–40.

[92] Swartz, DE, Wittenberg RH, Shea M, White AA, 3rd, Hayes WC. Physical and mechanical properties of calf lumbosacral trabecular bone. J Biomech 1991;24: 1059–68.

[93] Mosekilde, L, Kragstrup J, Richards A. Compressive strength, ash weight, and volume of vertebral trabecular bone in experimental fluorosis in pigs. Calcif Tissue Int 1987;40: 318–22. DOI: 10.1007/BF02556693

[94] Mitton, D, Rumelhart C, Hans D, Meunier PJ. The effects of density and test conditions on measured compression and shear strength of cancellous bone from the lumbar vertebrae of ewes. Med Eng Phys 1997;19: 464–74.

[95] Geusens, P, Boonen S, Nijs J, Jiang Y, Lowet G, Van Auderkercke R, Huyghe C, Caulin F, Very JM, Dequeker J, Van der Perre G. Effect of salmon calcitonin on femoral bone quality in adult ovariectomized ewes. Calcif Tissue Int 1996;59: 315–20. DOI: 10.1007/s002239900132

[96] Lim, TH, Hong JH. Poroelastic properties of bovine vertebral trabecular bone. J Orthop Res 2000;18: 671–7. DOI: 10.1002/jor.1100180421

[97] Jorgensen, CS, Kundu T. Measurement of material elastic constants of trabecular bone: a micromechanical analytic study using a 1 GHz acoustic microscope. J Orthop Res 2002;20: 151–8. DOI: 10.1016/S0736-0266(01)00061-4

[98] Courtney, AC, Hayes WC, Gibson LJ. Age-related differences in post-yield damage in human cortical bone. Experiment and model. J Biomech 1996;29: 1463–71. DOI: 10.1016/0021-9290(96)84542-8 4.3.2

[99] Joo, W, Jepsen KJ, Davy DT. The effect of recovery time and test conditions on viscoelastic measures of tensile damage in cortical bone. J Biomech 2007;40: 2731–7. DOI: 10.1016/j.jbiomech.2007.01.005 4.3.2

[100] Nyman, JS, Roy A, Reyes MJ, Wang X. Mechanical behavior of human cortical bone in cycles of advancing tensile strain for two age groups. J Biomed Mater Res A 2009;89: 521–9. DOI: 10.1002/jbm.a.31974 4.3.2

[101] Leng, H, Dong XN, Wang X. Progressive post-yield behavior of human cortical bone in compression for middle-aged and elderly groups. J Biomech 2009;42: 491–7. DOI: 10.1016/j.jbiomech.2008.11.016 4.3.2

[102] Keaveny, TM, Wachtel EF, Zadesky SP, Arramon YP. Application of the Tsai-Wu quadratic multiaxial failure criterion to bovine trabecular bone. J Biomech Eng 1999;121: 99–107. DOI: 10.1115/1.2798051 4.3.2

[103] Galante, J, Rostoker W, Ray RD. Physical properties of trabecular bone. Calcif Tissue Res 1970;5: 236–46. DOI: 10.1007/BF02017552

[104] Kaplan, SJ, Hayes WC, Stone JL, Beaupre GS. Tensile strength of bovine trabecular bone. J Biomech 1985;18: 723–7.

[105] Lucksanasombool, P, Higgs WA, Higgs RJ, Swain MV. Fracture toughness of bovine bone: influence of orientation and storage media. Biomaterials 2001;22: 3127–32. DOI: 10.1016/S0142-9612(01)00062-X 4.3.3

[106] Wang, X, Agrawal CM. Fracture toughness of bone using a compact sandwich specimen: effects of sampling sites and crack orientations. J Biomed Mater Res 1996;33: 13–21. DOI: 10.1002/(SICI)1097-4636(199621)33:1%3C13::AID-JBM3%3E3.0.CO;2-P 4.3.3

[107] Malik, CL, Stover SM, Martin RB, Gibeling JC. Equine cortical bone exhibits rising R-curve fracture mechanics. J Biomech 2003;36: 191–8. DOI: 10.1016/S0021-9290(02)00362-7 4.3.3

[108] Vashishth, D, Behiri JC, Bonfield W. Crack growth resistance in cortical bone: Concept of microcrack toughening. Journal of Biomechanics 1997;30: 763–769. DOI: 10.1016/S0021-9290(97)00029-8 4.3.3

[109] Feng, Z, Rho J, Han S, I Z. Orientation and loading condition dependence of fracture toughness in cortical bone. In International Conference on Advanced Materials. Beijing, China: Elsevier Sequoia; 1999. p. 41–46. DOI: 10.1016/S0928-4931(00)00142-9 4.3.3

[110] Norman, TL, Nivargikar SV, Burr DB. Resistance to crack growth in human cortical bone is greater in shear than in tension. J Biomech 1996;29: 1023–31. DOI: 10.1016/0021-9290(96)00009-7 4.3.3

[111] Corondan, G, Rottenberg N, Birzu S, Fitarau V, Kun G. Study on the arrangement of the fibrous tissue of bone. Acta Anat (Basel) 1967;67: 95–112. 4.3.3

[112] Barth, RW, Williams JL, Kaplan FS. Osteon morphometry in females with femoral neck fractures. Clin Orthop Relat Res 1992: 178–86. 4.3.3

[113] Squillante, RG, Williams JL. Videodensitometry of osteons in females with femoral neck fractures. Calcif Tissue Int 1993;52: 273–7. DOI: 10.1007/BF00296651 4.3.3

[114] Moyle, DD, Welborn JW, Cooke FW. Work to fracture of canine femoral bone. J Biomech 1978;11: 435–40. 4.3.3

[115] Saha, S, Hayes WC. Relations between tensile impact properties and microstructure of compact bone. Calcif Tissue Res 1977;24: 65–72. DOI: 10.1007/BF02223298 4.3.3

[116] Norman, TL. Microdamage of human cortical bone: incidence and morphology in long bones. Bone 1997;20: 375–9. DOI: 10.1016/S8756-3282(97)00004-5 4.3.3

[117] Schaffler, MB, Choi K, Milgrom C. Aging and matrix microdamage accumulation in human compact bone. Bone 1995;17: 521–525. DOI: 10.1016/8756-3282(95)00370-3 4.3.3

[118] Yeni, YN, Norman TL. Calculation of porosity and osteonal cement line effects on the effective fracture toughness of cortical bone in longitudinal crack growth. J Biomed Mater Res 2000;51: 504–9. 4.3.3

[119] Krajcinovic, D, Trafimow J, Sumarac D. Simple constitutive model for a cortical bone. J Biomech 1987;20: 779–84. 4.3.3

[120] Behiri, JC, Bonfield W. Fracture mechanics of bone–the effects of density, specimen thickness and crack velocity on longitudinal fracture. J Biomech 1984;17: 25–34. 4.3.3

[121] Currey, JD. Changes in the impact energy absorption of bone with age. J Biomech 1979;12: 459–69. 4.3.3

[122] Hiller, LP, Stover SM, Gibson VA, Gibeling JC, Prater CS, Hazelwood SJ, Yeh OC, Martin RB. Osteon pullout in the equine third metacarpal bone: effects of ex vivo fatigue. J Orthop Res 2003;21: 481–8. DOI: 10.1016/S0736-0266(02)00232-2 4.3.3

[123] Moyle, DD, Gavens AJ. Fracture properties of bovine tibial bone. J Biomech 1986;19: 919–27. 4.3.3

[124] Yeni, YN, Brown CU, Wang Z, Norman TL. The influence of bone morphology on fracture toughness of the human femur and tibia. Bone 1997;21: 453–9. DOI: 10.1016/S8756-3282(97)00173-7 4.3.3

[125] Bajaj, D, Arola DD. On the R-curve behavior of human tooth enamel. Biomaterials 2009;30: 4037–46. DOI: 10.1016/j.biomaterials.2009.04.017

[126] El, Mowafy OM, Watts DC. Fracture toughness of human dentin. J Dent Res 1986;65: 677–81.

[127] Zioupos, P, Currey JD. Changes in the stiffness, strength, and toughness of human cortical bone with age. Bone 1998;22: 57–66.

[128] Norman, TL, Vashishth D, Burr DB. Fracture toughness of human bone under tension. J Biomech 1995;28: 309–20. DOI: 10.1016/0021-9290(94)00069-G

[129] Nalla, RK, Kruzic JJ, Kinney JH, Ritchie RO. Effect of aging on the toughness of human cortical bone: evaluation by R-curves. Bone 2004;35: 1240–6. DOI: 10.1016/j.bone.2004.07.016

[130] Wright, TM, Hayes WC. Fracture mechanics parameters for compact bone–effects of density and specimen thickness. J Biomech 1977;10: 419–30.

[131] Melvin, JW E, FG. Biomechanics Symposium. In ASME; 1973. p. 87–88.

[132] Yan, J, Mecholsky JJ, Jr., Clifton KB. How tough is bone? Application of elastic-plastic fracture mechanics to bone. Bone 2007;40: 479–84. DOI: 10.1016/j.bone.2006.08.013

[133] Bonfield, W, Grynpas MD, Young RJ. Crack velocity and the fracture of bone. J Biomech 1978;11: 473–9. DOI: 10.1016/0021-9290(78)90058-1

[134] Yan, J, Clifton KB, Mecholsky JJ, Jr., Reep RL. Fracture toughness of manatee rib and bovine femur using a chevron-notched beam test. J Biomech 2006;39: 1066–74. DOI: 10.1016/j.jbiomech.2005.02.016

[135] Robertson, DM, Robertson D, Barrett CR. Fracture toughness, critical crack length and plastic zone size in bone. J Biomech 1978;11: 359–64.

[136] Yan, J, Daga A, Kumar R, Mecholsky JJ. Fracture toughness and work of fracture of hydrated, dehydrated, and ashed bovine bone. J Biomech 2008;41: 1929–36. DOI: 10.1016/j.jbiomech.2008.03.037

[137] Bonfield, W, Datta PK. Fracture toughness of compact bone. J Biomech 1976;9: 131–4.

[138] Phelps, JB, Hubbard GB, Wang X, Agrawal CM. Microstructural heterogeneity and the fracture toughness of bone. J Biomed Mater Res 2000;51: 735–41. DOI: 10.1002/1097-4636(20000915)51:4%3C735::AID-JBM23%3E3.0.CO;2-G

[139] Zimmermann, EA, Launey ME, Barth HD, Ritchie RO. Mixed-mode fracture of human cortical bone. Biomaterials 2009;30: 5877–84. DOI: 10.1016/j.biomaterials.2009.06.017

[140] Currey, JD, Brear K, Zioupos P. The effects of ageing and changes in mineral content in degrading the toughness of human femora. J Biomech 1996;29: 257–60. DOI: 10.1016/0021-9290(95)00048-8

[141] Cook, RB, Zioupos P. The fracture toughness of cancellous bone. J Biomech 2009;42: 2054–60. DOI: 10.1016/j.jbiomech.2009.06.001 4.3.3

[142] Panjabi, MM, White AA, Southwick WO. Mechanical properties of bone as a function of rate of deformation. J Bone Joint Surg Am 1973;55: 322–30. 4.3.4

[143] Hansen, U, Zioupos P, Simpson R, Currey JD, Hynd D. The effect of strain rate on the mechanical properties of human cortical bone. J Biomech Eng 2008;130: 011011. DOI: 10.1115/1.2838032 4.3.4

[144] Zioupos, P, Hansen U, Currey JD. Microcracking damage and the fracture process in relation to strain rate in human cortical bone tensile failure. J Biomech 2008;41: 2932–9. DOI: 10.1016/j.jbiomech.2008.07.025 4.3.4

[145] Crowninshield, R, Pope M. The response of compact bone in tension at various strain rates. Annals of Biomedical Engineering 1974;2: 217–225. DOI: 10.1007/BF02368492 4.3.4

[146] Bird, F, Becker H, Healer J, Messer M. Experimental determination of the mechanical properties of bone. Aerosp Med 1968;39: 44–8. 4.3.4

[147] Yamashita, J, Furman BR, Rawls HR, Wang X, Agrawal CM. The use of dynamic mechanical analysis to assess the viscoelastic properties of human cortical bone. J Biomed Mater Res 2001;58: 47–53. 4.3.4

[148] Yamashita, J, Li X, Furman BR, Rawls HR, Wang X, Agrawal CM. Collagen and bone viscoelasticity: a dynamic mechanical analysis. J Biomed Mater Res 2002;63: 31–6. 4.3.4

[149] Linde, F. Elastic and viscoelastic properties of trabecular bone by a compression testing approach. Dan Med Bull 1994;41: 119–38. 4.3.4

[150] Swanson, SA, Freeman MA. Is bone hydraulically strengthened? Med Biol Eng 1966;4: 433–8. DOI: 10.1007/BF02476165 4.3.4

[151] Pugh, JW, Rose RM, Radin EL. Elastic and viscoelastic properties of trabecular bone: dependence on structure. J Biomech 1973;6: 475–85. 4.3.4

[152] Kafka, V. On hydraulic strengthening of bones. Biorheology 1983;20: 789–93. 4.3.4

[153] Brown, TD, Ferguson AB, Jr. Mechanical property distributions in the cancellous bone of the human proximal femur. Acta Orthop Scand 1980;51: 429–37. DOI: 10.3109/17453678008990819 4.3.4

[154] Linde, F, Norgaard P, Hvid I, Odgaard A, Soballe K. Mechanical properties of trabecular bone. Dependency on strain rate. J Biomech 1991;24: 803–9. 4.3.4

[155] Zioupos, P, X TW, Currey JD. The accumulation of fatigue microdamage in human cortical bone of two different ages in vitro. Clin Biomech (Bristol, Avon) 1996;11: 365–375. DOI: 10.1016/0268-0033(96)00010-1 4.3.5

[156] Zioupos, P, Currey JD, Casinos A. Tensile Fatigue in Bone: Are Cycles-, or Time to Failure, or Both, Important? Journal of Theoretical Biology 2001;210: 389–399. DOI: 10.1006/jtbi.2001.2316 4.3.5

[157] Freeman, MA, Todd RC, Pirie CJ. The role of fatigue in the pathogenesis of senile femoral neck fractures. J Bone Joint Surg Br 1974;56-B: 698–702. 4.3.5

[158] Carter, DR, Hayes WC. Compact bone fatigue damage–I. Residual strength and stiffness. Journal of Biomechanics 1977;10: 325–337. 4.3.5

[159] Pidaparti, RM, Akyuz U, Naick PA, Burr DB. Fatigue data analysis of canine femurs under four-point bending. Biomed Mater Eng 2000;10: 43–50. 4.3.5

[160] Pattin, CA, Caler WE, Carter DR. Cyclic mechanical property degradation during fatigue loading of cortical bone. Journal of Biomechanics 1996;29: 69–79. DOI: 10.1016/0021-9290(94)00156-1 4.3.5

[161] Linde, F, Hvid I. Stiffness behaviour of trabecular bone specimens. J Biomech 1987;20: 83–9. 4.3.5

[162] Michel, MC, Guo XD, Gibson LJ, McMahon TA, Hayes WC. Compressive fatigue behavior of bovine trabecular bone. J Biomech 1993;26: 453–63. 4.3.5

[163] Bowman, SM, Guo XE, Cheng DW, Keaveny TM, Gibson LJ, Hayes WC, McMahon TA. Creep contributes to the fatigue behavior of bovine trabecular bone. J Biomech Eng 1998;120: 647–54. DOI: 10.1115/1.2834757 4.3.5

[164] Haddock, SM, Yeh OC, Mummaneni PV, Rosenberg WS, Keaveny TM. Similarity in the fatigue behavior of trabecular bone across site and species. J Biomech 2004;37: 181–7. DOI: 10.1016/S0021-9290(03)00245-8 4.3.5

[165] Caler, WE, Carter DR. Bone creep-fatigue damage accumulation. J Biomech 1989;22: 625–35. 4.3.5

[166] Ascenzi, A, E B. The tensile properties of single osteons. Anatomical Record 1967;158: 375–86. DOI: 10.1002/ar.1091580403 4.4.1, 4.4.2

[167] Ascenzi, A, E B. The compressive properties of single osteons. Anatomical Record 1968;161: 377–91. DOI: 10.1002/ar.1091610309 4.4.1, 4.4.2

[168] Ascenzi, A, P B, A B. The torsional properties of single selected osteons. Journal of Biomechanics 1994;27: 875–84. 4.4.1, 4.4.2

[169] Ascenzi, A, P B, A B. The bending properties of single osteons. Journal of Biomechanics 1990;23: 763–71.

[170] Choi, K, Kuhn JL, Ciarelli MJ, Goldstein SA. The elastic moduli of human subchondral, trabecular, and cortical bone tissue and the size-dependency of cortical bone modulus. J Biomech 1990;23: 1103–13. DOI: 10.1016/0021-9290(90)90003-L 4.4.1

[171] Choi, K, Goldstein SA. A comparison of the fatigue behavior of human trabecular and cortical bone tissue. J Biomech 1992;25: 1371–81. DOI: 10.1016/0021-9290(92)90051-2 4.4.1

[172] Jepsen, KJ, Schaffler MB, Kuhn JL, Goulet RW, Bonadio J, Goldstein SA. Type I collagen mutation alters the strength and fatigue behavior of Mov13 cortical tissue. Journal of Biomechanics 1997;30: 1141–1147. DOI: 10.1016/S0021-9290(97)00088-2

[173] Hernandez, CJ, Tang SY, Baumbach BM, Hwu PB, Sakkee AN, van der Ham F, De-Groot J, Bank RA, Keaveny TM. Trabecular microfracture and the influence of pyridinium and non-enzymatic glycation-mediated collagen cross-links. Bone 2005;37: 825–32. DOI: 10.1016/j.bone.2005.07.019 4.4.2

[174] Rho, JY, Tsui TY, Pharr GM. Elastic properties of human cortical and trabecular lamellar bone measured by nanoindentation. Biomaterials 1997;18: 1325–30. DOI: 10.1016/S0142-9612(97)00073-2 4.5

[175] Rho, JY, Zioupos P, Currey JD, Pharr GM. Variations in the individual thick lamellar properties within osteons by nanoindentation. Bone 1999;25: 295–300. 4.5

[176] Martin, B. A finite element model for mineral transport into newly-formed bone. Transactions of the Orthopaedic Research Society 1993;18: 80. 4.5

[177] Paschalis, EP, DiCarlo E, Betts F, Sherman P, Mendelsohn R, Boskey AL. FTIR microspectroscopic analysis of human osteonal bone. Calcif Tissue Int 1996;59: 480–7. DOI: 10.1007/BF00369214 4.5

[178] Tai, K, Qi HJ, Ortiz C. Effect of mineral content on the nanoindentation properties and nanoscale deformation mechanisms of bovine tibial cortical bone. J Mater Sci Mater Med 2005;16: 947–59. DOI: 10.1007/s10856-005-4429-9 4.5

[179] Turner, CH, Rho J, Takano Y, Tsui TY, Pharr GM. The elastic properties of trabecular and cortical bone tissues are similar: results from two microscopic measurement techniques. J Biomech 1999;32: 437–41. DOI: 10.1016/S0021-9290(98)00177-8 4.5

[180] Rho, JY, Currey JD, Zioupos P, Pharr GM. The anisotropic Young's modulus of equine secondary osteones and interstitial bone determined by nanoindentation. 2001. 4.5

[181] Franzoso, G, Zysset PK. Elastic anisotropy of human cortical bone secondary osteons measured by nanoindentation. J Biomech Eng 2009;131: 021001. DOI: 10.1115/1.3005162 4.5

[182] Goodwin, KJ, Sharkey NA. Material properties of interstitial lamellae reflect local strain environments. J Orthop Res 2002;20: 600–6. DOI: 10.1016/S0736-0266(01)00152-8 4.5

[183] Hengsberger, S, Kulik A, Zysset P. Nanoindentation discriminates the elastic properties of individual human bone lamellae under dry and physiological conditions. Bone 2002;30: 178–84. 4.5

[184] Hoffler, CE, Guo XE, Zysset PK, Goldstein SA. An application of nanoindentation technique to measure bone tissue Lamellae properties. J Biomech Eng 2005;127: 1046–53. DOI: 10.1115/1.2073671 4.5

[185] Hengsberger, S, Kulik A, Zysset P. A combined atomic force microscopy and nanoindentation technique to investigate the elastic properties of bone structural units. Eur Cell Mater 2001;1: 12–7. 4.5

[186] Mulder, L, Koolstra JH, den Toonder JM, van Eijden TM. Intratrabecular distribution of tissue stiffness and mineralization in developing trabecular bone. Bone 2007;41: 256–65. DOI: 10.1016/j.bone.2007.04.188 4.5

[187] Donnelly, E, Baker SP, Boskey AL, van der Meulen MC. Effects of surface roughness and maximum load on the mechanical properties of cancellous bone measured by nanoindentation. J Biomed Mater Res A 2006;77: 426–35.

[188] Viswanath, B, Raghavan R, Ramamurty U, Ravishankar N. Mechanical properties and anisotropy in hydroxyapatite single crystals. Scripta Materialia 2007;57: 361–364. DOI: 10.1016/j.scriptamat.2007.04.027 4.6

[189] Saber-Samandari, S, Gross KA. Micromechanical properties of single crystal hydroxyapatite by nanoindentation. Acta Biomaterialia 2009;5: 2206–2212. DOI: 10.1016/j.actbio.2009.02.009 4.6

[190] Shen, ZL, Dodge MR, Kahn H, Ballarini R, Eppell SJ. Stress-strain experiments on individual collagen fibrils. Biophys J 2008;95: 3956–63. DOI: 10.1529/biophysj.107.124602 4.6

[191] Balooch, M, Habelitz S, Kinney JH, Marshall SJ, Marshall GW. Mechanical properties of mineralized collagen fibrils as influenced by demineralization. Journal of Structural Biology 2008;162: 404–410. DOI: 10.1016/j.jsb.2008.02.010 4.6

[192] Balooch, M, Wu-Magidi IC, Balazs A, Lundkvist AS, Marshall SJ, Marshall GW, Siekhaus WJ, Kinney JH. Viscoelastic properties of demineralized human dentin measured in water with atomic force microscope (AFM)-based indentation. J Biomed Mater Res 1998;40: 539–44.

DOI: 10.1002/(SICI)1097-4636(19980615)40:4%3C539::AID-JBM4%3E3.0.CO;2-G
4.6

[193] Lanyon, LE, Hampson WG, Goodship AE, Shah JS. Bone deformation recorded in vivo from strain gauges attached to the human tibial shaft. Acta Orthop Scand 1975;46: 256–68. DOI: 10.3109/17453677508989216 4.7

[194] Rubin, CT, Lanyon LE. Limb mechanics as a function of speed and gait: a study of functional strains in the radius and tibia of horse and dog. J Exp Biol 1982;101: 187–211. 4.7

[195] Burr, DB, Milgrom C, Fyhrie D, Forwood M, Nyska M, Finestone A, Hoshaw S, Saiag E, Simkin A. In vivo measurement of human tibial strains during vigorous activity. Bone 1996;18: 405–10. DOI: 10.1016/8756-3282(96)00028-2 4.7

[196] Lanyon, LE, Rubin CT. Static vs dynamic loads as an influence on bone remodelling. J Biomech 1984;17: 897–905. 4.7

[197] Aamodt, A, Lund-Larsen J, Eine J, Andersen E, Benum P, Husby OS. In vivo measurements show tensile axial strain in the proximal lateral aspect of the human femur. J Orthop Res 1997;15: 927–31. DOI: 10.1002/jor.1100150620 4.7

[198] Turner, AS, Mills EJ, Gabel AA. In vivo measurement of bone strain in the horse. Am J Vet Res 1975;36: 1573–9. 4.7

[199] Lanyon, LE, Smith RN. Bone strain in the tibia during normal quadrupedal locomotion. Acta Orthop Scand 1970;41: 238–48. DOI: 10.3109/17453677008991511 4.7

[200] Lanyon, LE. Analysis of surface bone strain in the calcaneus of sheep during normal loco-motion. Strain analysis of the calcaneus. J Biomech 1973;6: 41–9. 4.7

[201] Biewener, AA, Taylor CR. Bone strain: a determinant of gait and speed? J Exp Biol 1986;123: 383–400. 4.7

[202] Lanyon, LE, Baggott DG. Mechanical function as an influence on the structure and form of bone. J Bone Joint Surg Br 1976;58-B: 436–43. 4.7

[203] Manley, PA, Schatzker J, Sumner-Smith G. Evaluation of tension and compression forces in the canine femur in vivo. Arch Orthop Trauma Surg 1982;99: 213–6. DOI: 10.1007/BF00379211 4.7

[204] Szivek, JA, Johnson EM, Magee FP. In vivo strain analysis of the greyhound femoral diaphysis. J Invest Surg 1992;5: 91–108. DOI: 10.3109/08941939209012426 4.7

[205] O'Connor, JA, Lanyon LE, MacFie H. The influence of strain rate on adaptive bone remodelling. J Biomech 1982;15: 767–81. DOI: 10.1016/0021-9290(82)90092-6

[206] Lanyon, LE, Bourn S. The influence of mechanical function on the development and remodeling of the tibia. An experimental study in sheep. J Bone Joint Surg Am 1979;61: 263–73.

[207] Goodship, AE, Lanyon LE, McFie H. Functional adaptation of bone to increased stress. An experimental study. J Bone Joint Surg Am 1979;61: 539–46.

[208] Demes, B, Stern JT, Jr., Hausman MR, Larson SG, McLeod KJ, Rubin CT. Patterns of strain in the macaque ulna during functional activity. Am J Phys Anthropol 1998;106: 87–100. DOI: 10.1002/(SICI)1096-8644(199805)106:1%3C87::AID-AJPA6%3E3.0.CO;2-A

[209] Swartz, SM, Bertram JE, Biewener AA. Telemetered in vivo strain analysis of locomotor mechanics of brachiating gibbons. Nature 1989;342: 270–2. DOI: 10.1038/342270a0

[210] Keller, TS, Spengler DM. Regulation of bone stress and strain in the immature and mature rat femur. J Biomech 1989;22: 1115–27.

[211] Mosley, JR, March BM, Lynch J, Lanyon LE. Strain magnitude related changes in whole bone architecture in growing rats. Bone 1997;20: 191–8. DOI: 10.1016/S8756-3282(96)00385-7

[212] Biewener, AA, Fazzalari NL, Konieczynski DD, Baudinette RV. Adaptive changes in trabecular architecture in relation to functional strain patterns and disuse. Bone 1996;19: 1–8. DOI: 10.1016/8756-3282(96)00116-0

[213] Judex, S, Gross TS, Zernicke RF. Strain gradients correlate with sites of exercise-induced bone-forming surfaces in the adult skeleton. J Bone Miner Res 1997;12:1737–45.

CHAPTER 5

Structure and Properties of Scaffolds for Bone Tissue Regeneration

CHAPTER SUMMARY

The mechanical properties of scaffolds ideally should match those of the bone that is defective and needing repair. This is a major challenge for both the orthopaedic and biomaterial research communities because sufficient porosity is a pre-requisite to achieve osteogenesis, and this porosity weakens the scaffold. Polymers are popular choice for scaffold designs because their mechanical and degradation properties are 'tunable' via changes in molecular weight, crosslinking, and micro-architecture. The inclusion of bioceramics into polymer scaffolds increases the mechanical strength. Having similar constituents to bone, bioceramics also support bone formation and are osteoconductive. They do not however degrade as quickly as polymers. With degradation, there is a loss in the mechanical properties. Therefore, rate of new bone formation ideally should match the degradation rate, thereby maintaining the overall mechanical integrity of the scaffold as it repairs a bone defect. To assist this process, various morphogenetic proteins or growth factors as well as stromal or stem cells can be included into the scaffold. The success of these osteoinductive approaches depend on the release kinetics of the scaffold, and thus is a function of the micro-architecture, resorption characteristics, and mechanical behavior of the scaffold.

5.1 INTRODUCTION

A major challenge in orthopaedics is the repair of bone tissue with an implant that is enduring, weight bearing, and requiring no revision surgery or at least minimally invasive surgery. Presently, there are not any commercially available products that can (1) support forces equal to or greater than body weight, (2) regenerate bone tissue, and (3) degrade into harmless by-products. Therefore, nearly all skeletal defects caused by complex fractures or bone metastasis necessitate the use of metal implants (e.g., plates, screws, rods, cables) to stabilize the bone defect or to secure a bone allograft to the host bone, thereby facilitating tissue regeneration and repair. Alternatively, metal implants replace the defective bone completely (e.g., prosthesis). While metal implants provide successful outcomes in that patients return to daily activities with minimal pain, there are a percentage of cases that result in costly non-unions (i.e., bone does not heal) [1], and virtually all metal implants must be removed or replaced at some point [2].

Scaffolds provide a unique strategy to achieve the trifecta of enduring bone repair, which is the ability to degrade or resorb while bearing physiologic loads and regenerating tissue. As previously mentioned in Chapter 1, they provide a structure that fills a defect in the bone, allows for the infiltration of endogenous cells or the release of exogenous cells, diffusion of nutrients and waste products, and the delivery of therapeutics such as antibiotics or growth factors. With regards to bone regeneration, scaffolds must be osteoconductive, osteoinductive, or both. That is, they must either support bone cell functions (osteoclast-mediated resorption and osteoblast-mediated bone formation) or stimulate bone formation across a gap. To achieve such attributes, scaffold design depends on the appropriate selection of biomaterials and architecture.

Biomaterials fall under three general classifications: polymers, ceramics, and metals. All 3 types of biomaterials plus composites have been developed into scaffolds for bone regeneration. Polymers can be further classified as being either synthetic or naturally derived, both of which have been developed for bone repair. The mechanical function of the tissue requiring regeneration dictates the choice of biomaterial(s) and scaffold architecture.

In this chapter, we describe (1) determinants of mechanical properties for 4 general types of biomaterials used in scaffolds, namely polymers, bioactive ceramics, titanium, and polymer/ceramic composites; (2) structural requirements for functional scaffolds (e.g., porosity, permeability, etc.); and (3) balance between scaffold degradation and tissue regeneration.

5.2   SYNTHETIC POLYMERS

5.2.1   DETERMINANTS OF THE MECHANICAL PROPERTIES OF SYNTHETIC POLYMERS

Polymers are commonly used as a biomaterial because they can be engineered to have a wide range of mechanical properties (kPa to GPa) and thus can be applied to a wide variety of pathological conditions (e.g., from arteriosclerosis to osteoarthritis). In addition, many polymers are biodegradable and biocompatible. The functional unit or building block of the polymer is a chemical entity known as a mer that is serially linked via carbon-carbon bonds (Figure 5.1). Synthetic polymers may have more than one mer (copolymer), and in naturally derived tissues, the mer is not just one compound but also one of the various amino acids that are linked into a peptide. Certain mers undergo hydrolysis with and without enzymatic activity and thus degrade in the aqueous environment of the body, whereas sequences of naturally derived polymers are recognized by enzymes and thus undergo proteolysis [3].

Ideally, the scaffold should have the strength and stiffness of the bone tissue that it is trying to regenerate, and as previously discussed, bone strength can range from 20 MPa (trabecular) to 200 MPa (cortical). One advantage of synthetic polymers then is the ability to 'tune' the mechanical properties to match the desired function. There is a trade-off however between strength and the ability to degrade. Generally, speaking, strong polymers degrade slowly, whereas, highly degradable polymers are weak. The mechanical properties depend on the size of the mer, the arrangement of

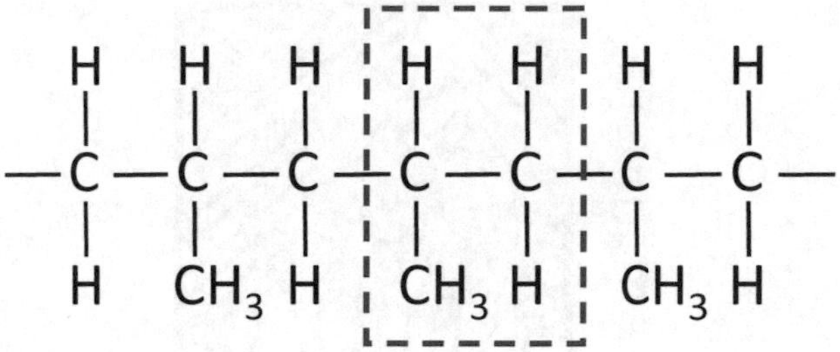

Figure 5.1: The structural unit of a polymer is a repeat of a chemical entity or mer along serially linked carbon-carbon bonds [4].

the side chains of the mer, the functionality of the mer, and synthesis techniques because all these factors dictate the molecular weight (MW) and crystallinity of the polymer [3].

Polymers become stronger, stiffer, tougher, and more wear resistant with an increase in MW (Table 5.1). The degree of polymerization or the number of mers serially linked together dictates the MW. Here, synthesis is important. Addition polymerization that uses a free radical initiator in a controlled environment can generate long chains of carbon-carbon bonds, especially for mers with a simple chemical structure (e.g., ethylene). In contrast, condensation polymerization generates short chains because the chemical species (e.g., amine + carboxylic acid) in the step-wise reactions have limited mobility. Short chains (low MW) more readily slide relative to one another than long chains (high MW), which typically become entangled with one another. This decrease in relative movement with increasing chain length causes the increase in rigidity and strength of the polymer [3,4].

Table 5.1: In general, there is a trade-off between an increase in the mechanical properties and a decrease in the degradation rate with an increase in MW [4].

Molecular weight of PLA	% Mass loss
89,000	21
199,000	15
266,000	10
294,000	7

Increasing the crystallinity or density of the polymer also increases the rigidity, strength, and wear resistance of the implant or scaffold [5]. Crystallinity characterizes the fraction or percentage of the polymer chains that are arranged in an orderly fashion with the carbon-carbon backbone folding over itself into a lamellar array (Figure 5.2). Otherwise, the chain is amorphous that is randomly twisting and turning around neighboring chains like a bowl of spaghetti. The movement of the crystalline chains is considerably less than the movement of non-crystalline chains [3]. Factors

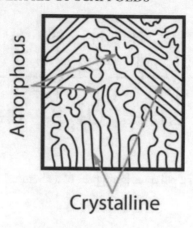

Figure 5.2: There are typically two phases in polymers: one in which the carbon-carbon chains fold over one another to produce an order array (crystalline) and another in which the chains twist and turn in a random fashion (amorphous).

affecting crystallinity include (1) side groups of the mer(R), (2) functionality of the mer, (3) arrangement or tacticity of the side groups along the chain, and (4) arrangement of the different mers in copolymers. Bulky side-groups hinder rotational freedom of the carbon-carbon backbone and thus lower crystallinity since the chain cannot readily fold over itself or interact with neighboring chains. Tri-functional mers can bond with three other mers and thus facilitate branching of the polymer chain. Branching also impedes the orderly arrangement of chains, although it does entangle the non-crystalline phase more so than bi-functional, linear chains. When the side groups are atactically arranged (Figure 5.3), the crystallinity is reduced because again neighboring polymer chains are less likely to come together than in the case of an isotactic or syndiotactic arrangement (Figure 5.3). Similarly, the random arrangement of differing mers in a block copolymer precludes orderly folding and packing of the chain into a lamellar structure [3].

Crosslinking the polymer chains also impedes movement and facilitates crystalline arrangements. Thus, crosslinking can increase the strength and rigidity of polymers such as polyethylene [6] and polyanhydrides [7]. Since crosslinks are usually covalent between two carbons on neighboring chains, the chains are physically closer, thereby increasing density of the polymer.

The ability of polymer chains to slide relative to one another and unravel provides the mechanistic explanation for the relationship between the micro-structure and the mechanical properties of polymers. In a sense, when polymers deform, they have flow-like characteristics and are indeed a viscoelastic material. As is the case with bone, they exhibit creep under constant stress, stress relaxation under constant strain, and strain-dependent mechanical properties [3]. Not surprisingly then, varying crosslink concentration and crystallinity adjusts the viscoelastic properties, but achiev-

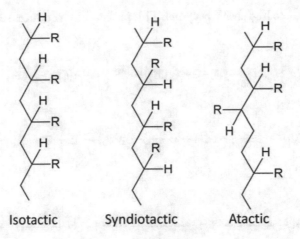

Isotactic   Syndiotactic   Atactic

Figure 5.3: The arrangement of side groups affects crystallinity and ultimately the mechanical properties.

ing the same viscoelastic behavior of bone is rather difficult without losing the ability to undergo bioresorption.

5.2.2 EXAMPLES OF SYNTHETIC POLYMERS FOR BONE REGENERATION

Although numerous synthetic polymers (e.g., polyfumarates [8], polyanhydrides [9]) have been developed into scaffolds for bone tissue engineering, the following descriptions are limited to polyesters and polyurethanes as examples. There is currently a polyester-based scaffold for bone regeneration (Osteofoam by BoneTec [10,11]), and polyurethanes are potentially promising scaffolds since they have the capability of being injectable and porous with tunable mechanical properties [12]. The following descriptions are primarily related to factors affecting mechanical properties, not other important issues such as biocompatibility.

Polyesters are one class of synthetic polymers currently being investigated for bone tissue regeneration and include poly (lactic acid) (PLA can be PLLA, PDLA, or PDLLA depending on chirality or handedness of compound or compounds), poly(glycolic acid) (PGA), poly(lactic-*co*-glycolic acid) (PLGA), poly(e-caprolactones) (PCL), and polycarbonate. Since they degrade in weeks to months, polyester-based scaffolds enable natural tissue to replace the implant in a process referred to as creeping substitution [13]. The central chemistry of these polymers involves reactions of α-hydroxy acids in which the reactive groups are on opposite ends of the chemical compound. Because polymerization occurs via condensation (i.e., involves the release of water), the carbon chains of polyesters are not particularly long. To produce an increase in the MW of polyesters, the α-hydroxy acid is converted into cyclic ester or what is referred to as a lactone [14]. Glycolide, lactide, and e-caprolactone (Figure 5.4) are examples of common lactones that can undergo ring-opening polymerization to produce high MW polymer. This process also produces semi-crystalline polymers

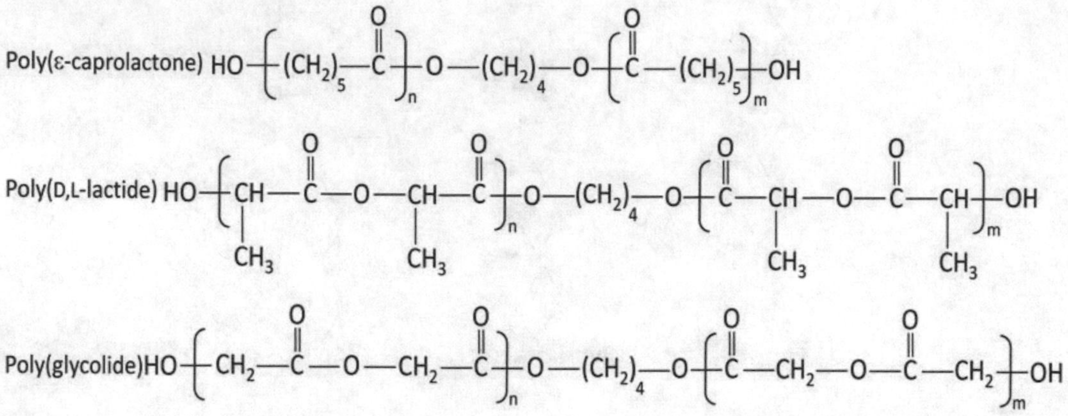

Figure 5.4: The chemical name and structure of common polyesters that have been developed into scaffolds for bone regeneration.

such that a slow rate of cooling during the reaction increases the time for the chains to move into an ordered structure. The maximum tensile and flexural strength of PLA is on the order of 80 MPa and 145 MPa, and, historically, these types of polymers have been used as absorbable implant devices (e.g., bone screws) [15].

The ability of polyester-based scaffolds to support bone cell attachment and proliferation is well established [16, 17]. In one of the earlier studies demonstrating the ability of polyesters to regenerate bone, Puelacher et al. implanted PGA-based scaffolds into critical sized defects (9 mm) that were created in the femur diaphysis of athymic rats (immune compromised) and found that when the scaffolds were cultured with periosteum-derived cells (bovine) *in vitro* prior to implantation, new bone formation bridged the gap after 12 weeks to a greater extent than when the scaffold alone (no cells) or no scaffold was implanted [18]. Note that titanium pins were used to stabilize the femur with large defects since the mechanical properties of PGA are not sufficient to withstand physiological forces. In a more recent study by Ge et al. using a rabbit model [19], PLGA-based scaffolds were implanted into the iliac crest, within the periosteum and separately within a bi-cortical bone defect, and subsequent histological analysis at 4, 12, and 24 weeks of healing revealed that scaffolds supported new bone formation over time. The novelty of this study was the use of 3D printing technology to generate scaffolds with strength comparable to cancellous bone and with a shape matching the bone defect.

The addition of polyethylene glycol (PEG) to polyesters produce hydrogels that are viscoelastic with low mechanical strength (in the kPa range) [20]. Nonetheless, they have been used to regenerate bone [21,22]. One attractive feature is that Arg-Gly-Asp (RGD) peptide sequence are readily grafted on to the hydrogels, and this sequence has been observed to promote osteoblast attachment and cell density that resulted in greater mineralization in an *in vitro* culture assay [23]. The success of hydrogels for repairing bone defects likely depends on identification of the right growth factors to accelerate osteogenesis or the right *in vitro* conditions to generate mineralization and increase strength prior to implantation.

Polyurethanes (PUR) are another class of polymers being developed as biodegradable scaffolds and synthetic grafts for bone regeneration. The reaction of isocyanates with a hydroxyl-functional molecules generates the urethane linkages and additional reactions with amines generate the urea linkages (Figure 5.5) [24]. In the presence of water, the isocyanates decompose into an amine and carbon dioxide, and the gas acts like a blowing agent creating a porous scaffold (Figure 5.6). The hydroxyl-functional molecules are viscous liquids called polyols with a polymer backbone (e.g., polyester) between the OH groups. To gain further control over the MW of the PUR, the polyol can react with excessive diisocyanate to generate prepolymers with a NCO group at the end. Varying the ratio of NCO to OH changes the properties of the PUR. To increase the MW and hence strength, the prepolymer typically is made to react with another polyol or polyamine to extend the chain. Since polyol and urethane segments are soft relative to the hard urea that is connected to the chain extender segments, the PUR has two microphases, which can be manipulated to generate scaffolds with a wide range of mechanical properties. Increasing the hard phase relative to the soft phase via selection of the molecules comprising the chain extenders increases the strength and stiffness of PUR [24].

Guelcher and colleagues performed in vitro cell culture experiments and demonstrated the potential for PUR to serve as scaffolds for bone regeneration [25,26]. Using tyramine and tyrosine moieties for chain extenders in order to promote the hard phase, biodegradable segmented PUR elastomers were synthesized such that varying the molecular weight of the PCL from 1100 to 2700 g/mol increased the melting temperature from 21 to 61 °C and increased the storage modulus from 52 MPa to 278 MPa at 37 °C. When bone marrow stromal cells were cultured on rigid scaffolds of these PUR scaffolds in osteogenic media, expressions of bone forming genes (e.g., alkaline phosphates, osteopontin, osteocalcin) were the same as those for stromal cells cultured on scaffolds made of a PLGA substrate [26]. In addition, these PUR-based scaffolds support attachment and proliferation of viable human osteoblast-like cells (MG-63) [25]. As for in vivo evidence that PUR can regenerate bone tissue, Gogolewski and Gorna found that when porous scaffolds (pore-to-volume ratio = 85 %) of biodegradable, segmented aliphatic polyurethanes were either implanted into biocortical defects created in the tuber coxae of estrogen-deficient, mature sheep [27] or monocortical defects in the iliac crest of healthy, intact sheep [28], newly formed, well mineralized bone tissue filled the defect to greater extent than when there was no implant (control). Increasing the amount of hydrophilic components increased the degree of mineral deposition, and addition of calcium-complexing agent improved bone healing. To identify new bone formation, the bone with the scaffold is fixed, embedded, and sectioned for histological staining (Figure 5.7).

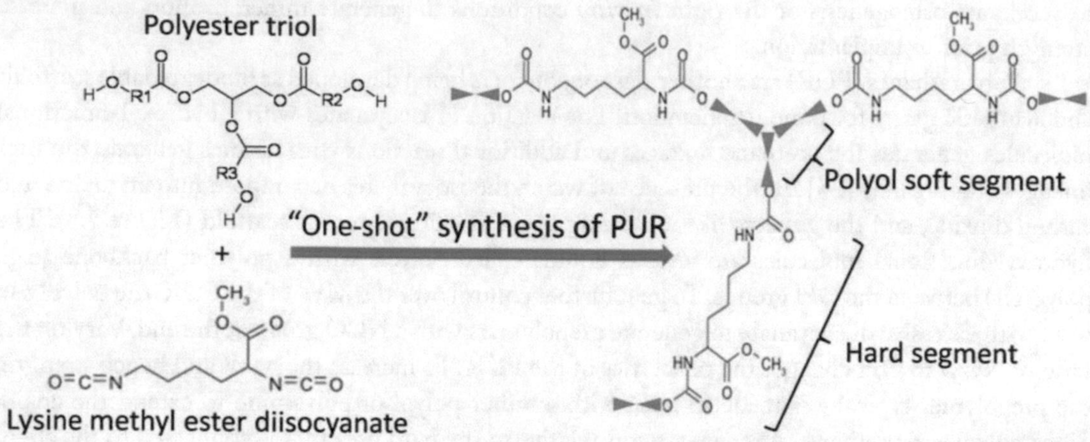

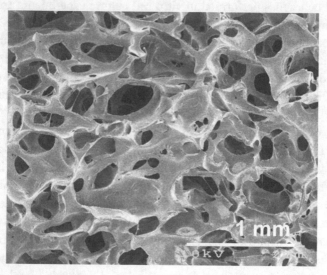

Figure 5.5: In the synthesis of polyurethanes, the reaction of diisocyanate with hydroxyl-functional molecule (polyol) generates polyurethane (PUR). The choices of the polyol and extenders dictate the strength of the polyurethane through effects on the relative fraction of the hard and soft segments.

Figure 5.6: Imaging a scaffold by scanning electron microscopy reveals the architecture of the porosity. SEM image courtesy of Dr. Scott Guelcher.

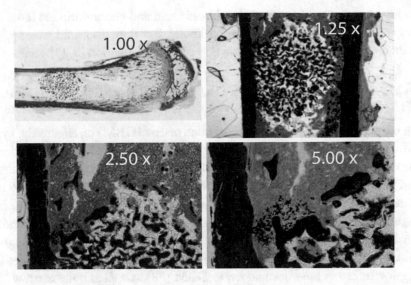

Figure 5.7: When a PUR scaffold was implanted into the tibia diaphysis of euthymic rats, there was new bone formation (black stain) after 3 weeks. Histological images courtesy of Dr. Scott Guelcher.

5.2.3   DETERMINANTS OF THE MECHANICAL PROPERTIES OF NATURALLY DERIVED POLYMERS

The factors affecting the mechanical properties of naturally derived polymers are similar to those discussed in previous chapters describing the composition and mechanical behavior of bone. However, when used as scaffolds to deliver growth factors or cells for bone repair, naturally derived polymers are initially demineralized unless being developed as a composite. Therefore, they are weaker and more compliant than mineralized bone allograft. When collagen content or density and crosslinking is increased, scaffolds become stronger [29]. The effect of crosslinking on collagen strength depends on immersion time and the agent being used (e.g., glutaraldehyde vs. grape seed extract) [30]. Also crosslinking heparin to Type I collagen can increase scaffold strength [31]. Moreover, there is considerable heterogeneity in the mechanical properties of demineralized bone without treatments, and this is likely due to differences in collagen structure and organization for a given anatomical site [32]. Typical values of modulus, ultimate strength and ultimate strain of collagen from bone are 613 ± 113 MPa, 61.5 ± 13.1 MPa, and $12.3\pm0.5\%$, respectively [33]. There is, however, a certain degree of anisotropy since demineralized bone is typically stronger when the applied load is aligned to the long axis of the majority of the collagen fibrils [34].

Other factors that affect naturally derived polymers include the presence of elastin, polysaccharides, hyaluronic acid, and chondrotin [5]. Although not as abundant in bone as in ligaments, elastin is primarily composed of non-polar amino acid residues and as such is highly elastic (e.g., springs back upon unloading). Polysaccharides are simple sugars that promote viscoelastic behavior since they readily bind water and cations. Hyaluronic acid and chondroitin are more complex than the simple sugars since they have a protein backbone, but they also hold water within collagen matrices and increase stiffness in compression and provide lubrication [5,35].

5.2.4 EXAMPLES OF NATURALLY DERIVED POLYMERS FOR BONE REGENERATION

One common natural polymer is demineralized bone matrix (DBM) or, effectively, Type I collagen. Being derived from biological tissue, DBM supports cellular attachment and proliferation. It also supports bone regeneration and is FDA approved for spine fusion. In the initial assessment, DBM was implanted into a circumferential defect of the diaphysis of the ulna with 82% of the rabbits exhibiting bone formation in the implant by 12 weeks [36]. The ability of DBM to facilitate bone formation depends on the source of bone in which demineralized intramembranous bone (the type found in the skull) improves the osteoconductivity of autogenous bone allograft (either of endochondral or intramembranous origin) as determined in a rabbit defect model [37]. Various recombinant growth factors have been added to DBM [38] as well as bone marrow [39] to hasten the rate of bone regeneration. Lastly, gamma irradiation lessens the osteoconductive potential of DBM, though this is minimized when performed on dry collagen at low temperatures ($< -40\,°C$ to lessen radiolysis) [40].

Chitosan is another naturally derived polymer being used in scaffolds to regenerate bone tissue. Derived from crustacean shells, it is a linear polysaccharide with a MW in the range of 300 to over 1000 kD. The factors affecting the functionality of chitosan include: MW, ionic strength (electrostatic attraction between the chain and negatively charged compounds such as proteoglycans and heparin), degree of deacetylation (30% to 95%), and stability [41,42]. Tensile strength of non-porous chitosan is 5-7 MPa, so scaffolds of this material are not generally weight bearing but are excellent vehicles for the delivery of cytokines or growth factors [42,43]. Even without growth factors, porous scaffolds of chitosan (i.e., sponges) have been shown to support the differentiation of primary osteoblasts from the calvaria of rats and the subsequent deposition of mineral [44]. As with other synthetic and naturally derived polymers, chitosan have been made into porous scaffolds and combined with calcium phosphate to increase strength [45].

5.3 BIOCERAMICS

5.3.1 DETERMINANTS OF THE MECHANICAL PROPERTIES OF CERAMICS

Initially, the application of ceramics to orthopaedics involved components of total hip arthroplasty because such ceramics as zirconia (ZrO_2) and alumina (Al_2O_3) are highly inert with excellent wear resistance. Such properties depend on the ionic bonding that occurs at the atomic level in that greater differences in the electronegativity between the elements of the ceramic generate greater electrostatic attractions (columbic force) and hence greater bonding strength and overall greater bulk modulus [3]. The condition of electroneutrality makes ceramics brittle since during plastic deformation a cation (positively charged) must bond with an anion (negatively charged), thereby increasing the inter-atomic distance that a dislocation (defect) must move. Ultimately, zirconia and alumina are strong (compressive strength of 900 MPa and 480 MPa, respectively) because they can be manufactured with small grains (0.5-1.8 microns) and low porosity [4].

5.3.2 EXAMPLES OF CERAMICS FOR BONE REGENERATION

Being dense and inert, the ceramics for prostheses do not biodegrade and facilitate bone formation. There are, however, bioceramics and bioactive glass-ceramics that are osteoconductive and bioresorbable [46]. These ceramics essentially have the same chemical constituents as bone (calcium and phosphate). Moreover, their crystallographic structure is similar, though chemical groups in bone are non-stoichiometric with substitutions of carbonate (typically CO_3 for PO_4 or what is known as Type B). The two most widely studied bioceramics as synthetic bone grafts are tricalcium phosphate (TCP or $Ca_2(PO_4)_2$) and hydroxyapatite (HA or $Ca_{10}(PO_4)_6(OH)_2$) [47]. The elastic modulus of these materials can vary from 4.0 GPa to 117 GPa depending on the porosity. Processing methods involving the sintering temperature and pressure influence density, crystalline structure (grain size), and porosity. That is, there is some control over the mechanical properties of bioceramics, though the range is not the same as polymers. Being ceramics though, TCP and HA are brittle by nature and are stronger in compression (around 294 MPa) than in bending (around 147 MPa) [48]. When developed into scaffolds however, the mechanical properties are much lower. Typically, these materials have been developed as injectable cements that harden *in vivo* with porosity. To improve the mechanical properties, HA nanofibers have been incorporated within a scaffold of TCP, but this technique still only produced a compressive strength of 9.8 ± 0.3 MPa and a fracture toughness of 1.72 ± 0.01 kN$\sqrt{}$m [49]. In addition, increasing the concentration of HA from 35% to 50% can increase the elastic modulus and yield strength, respectively, of porous scaffolds (pore size = 200-400 μm) made of calcium phosphate (Ca/P = 1.7) [50].

As the name implies, glass-ceramics possess crystallized glass (i.e., silca or SiO_2). The nucleation of the glass phase requires metallic precipitates and causes crystallization to reach 90% with grain sizes (0.1~1.0 μm) that are smaller than conventional ceramics [5]. By adding bioactive ceramics such as CaO and Na_2O at sufficient concentrations, the glass-ceramic bonds to bone [51] and can be osteoconductive when developed as granules to fill osseous lesions [52]. Glass-ceramics such as Bioglass (SiO_2-Na_2O-CaO-P_2O_5) are also brittle in nature with tensile strengths in the range of 40 MPa to 60 MPa [47]. The elastic modulus is, however, in the range of cortical bone (7~25 GPa). As such, bioactive glass-ceramics are used in non-weight bearing sites where they form strong bonds with hard and soft tissue through cellular activity. Interestingly, bioactive ceramic-glass typically generate greater bone formation than HA [53,54].

5.3.3 EXAMPLES OF CERAMIC-POLYMER COMPOSITES FOR BONE REGENERATION

To enhance osteoconductivity and mechanical strength, composites of polymers and bioactive ceramics are being developed into scaffolds for bone regeneration. Bone is after all a composite of a collagen matrix (i.e., a polymer) and a mineral apatite (i.e., ceramic). The ceramic strengthens the polymer [78], and vice versa, the polymer imparts ductility to the otherwise brittle ceramic. Examples of composites include the following: collagen–HA [55], chitosan–HA [56], PLA–HA [57], PLA–PEG–HA [58], HA–Bioglass [59], and collagen–PLA–hydroxyapatite [60].

Given their considerable promise to be weight bearing while generating bone tissue, these composites have been extensively reviewed [61–63]. Salient points from the literature follow. One, success of the composite sample depends on a processing technique that binds the ceramic to the polymer, for this affects the mechanical properties and degradation characteristics of the composite. Two, bioactive ceramics should not be entirely encased in the polymer matrix so that they can still act as osteoconducters. Three, the dissolution of the ceramic or salt phase can neutralize the acid byproducts of the polymer phase. Four, manipulating concentration of the components, temperature of reaction, and rate of cooling control the polymer microstructure and ultimately the characteristics of degradation and osteogenesis.

Polymer-ceramic composites do not always need to be made into a scaffold in order to regenerate bone. For example, Hasegawa et al. created intramedullary (IM) rods fabricated from composites incorporating 30–40 wt% HA and 60–70 wt% PLLA with bending strengths ranging from 260 – 280 MPa and elastic modulus ranging from 7.6 – 9.8 GPa. When they implanted the composite in the distal femur of rabbits, the composites partially remodeled and integrated with host tissue after 4 years [64]. In some of the rabbits that lived beyond 4 years, the rod was almost completely resorbed with new bone encasing the residual HA. Recently, carbon fibers have been introduced into PLA-HA composites to achieve a flexural strength of 430 MPa and a flexural modulus of 22 GPa at 15% HA [65]. In vitro degradation tests indicated that these properties decreased by 13.2% and 5.4%, respectively, after 3 months of incubation in saline at 37 °C. The degradation would likely be faster in vivo where cells (osteoclasts) could resorb the ceramic phase. The question then for this biomaterial, and all composites for that matter, is whether the new bone formation will match degradation rate, thereby maintaining strength.

5.4 METALS

5.4.1 DETERMINANTS OF THE MECHANICAL PROPERTIES OF METALS

Strength, modulus, and ductility of metals primarily depend on alloying elements, crystallographic structure and microstructure. Alloying is a type of solid solution in which substitution of point defects (e.g., carbon in the case of steel) are created within the lattice of the metal (e.g., iron) increasing the energy required to move dislocations. Dislocation glide or slip is the basis of plastic or permanent deformation of metals, and so mechanical properties also depend on the number of slip systems within the crystallographic structure (e.g., body centered cubic has 12 slip systems whereas hexagonal closed packed has 3). Typically, metals with small grains are stronger than those with large grains because the orientations of the slip systems are more random, varying from grain to grain, in the former microstructure than the latter and because plastic deformation does not occur until the resolved shear stresses in all neighboring grains exceed the critical level of the material [3]. With regards to metal scaffolds, the apparent strength depends on the porosity and architecture (i.e., fiber diameter) [66].

5.4.2   EXAMPLES OF METALS FOR BONE REGENERATION

Metals do not undergo hydrolysis or enzymatic degradation, and so they cannot be developed as a bioresorbable scaffold. Nonetheless, the use of titanium alloy as a porous coating on an implant surface or as a porous scaffold does facilitate bone formation. As one example, Li et al. implanted titanium alloy scaffolds between decorticated transverse processes of the lumbar spine of goats and found that increasing the porosity and pore size (controlled by a 3D fiber deposition technique) improved the osteoconductive properties [67]. Combining titanium meshes with coral, a naturally derived ceramic, to create a cell-seeding scaffold (i.e., bioreactor) also generated new bone formation after 2 months of implantation into the backs of athymic mice [68]. These scaffolds have also been combined with growth factors (BMP-2 and TGF-β1) with and without calcium phosphate and found to be highly osteoconductive [69, 70]. Titanium alloy possesses an oxidative layer that greatly impedes corrosion, and so even though this material will not undergo biodegradation as do biopolymers, the titanium mesh is likely to be long lasting given that the new bone deposition within its pores are interconnected, thereby imparting mechanical competence to the bone defect.

5.5   STRUCTURAL REQUIREMENTS FOR FUNCTIONAL SCAFFOLDS

A porous scaffold is an enhancer, if not a prerequisite, for bony in-growth *in vivo* [71]. Porosity of course weakens the implant, but having a pore size with a diameter of at least 300 μm facilitates the transport of nutrients and waste, the formation of vascular system, and ultimately osteogenesis [72]. In other words, interconnected pores are generally considered necessary to promote bone ingrowth into a polymeric scaffold, but the pre-existing pores significantly reduce the initial load-bearing properties of the device [61]. Unfortunately, scaffolds with small pores (high strength) tend to cause hypoxic conditions in the microenvironment favoring osteochondral formation, not osteogenesis [63]. With large pores (low structural strength), there is ample room for vascularization and deliver of the precursor cells of the osteoclast and osteoblast lineages. Shape of pore is also important because it can affect attachment and survival of the cells within the scaffold [10, 73], and tortuosity is another important characteristic of scaffold architecture since it can affect the delivery of growth factors to endogenous cells [61]. Aiding the process of osteogenesis is the bioresorption of polymer within the scaffold to create space for vasculature and new mineral deposition. The key is to match degradation rate of the new bone formation rate.

There are several strategies for hastening osteogenesis within the scaffold and thus the mechanical integrity of the bone defect. One is to seed stromal cells within the scaffold *in vitro* so that osteogenesis begins before implantation [74]. Mineralization, as it does for bone, increases the strength of the scaffold. There are numerous challenges with this approach including (1) the design of an efficient bioreactor to facilitate nutrient supply and waste removal [13], (2) rejection by the immune system, and (3) sterilization which can decrease the bioactivity. Placing the scaffold in flow perfusion system can enhance proliferation and differentiation of the stromal cells to deposit mineral [75]. Nonetheless, there are also issues of high manufacturing costs [76]. Encapsulating growth factors (e.g., BMP-2, VEGF, PDGF) or anabolic agents is another strategy to stimulate osteointegration of the implant, thereby counteracting the loss of strength of the polymer due to degradation.

5.6   THE BALANCE BETWEEN SCAFFOLD DEGRADATION AND TISSUE REGENERATION

The ideal scaffold degrades at the same rate that new bone tissue is formed (osteogenesis), thereby maintaining a constant mechanical strength across the skeletal defect. Although there are techniques available in the synthesis of biomaterials to control the degradation rate of polymers and there is enough information on the effects of scaffold porosity and its architecture on bone formation (as previously discussed), matching the rate of osteogenesis or osteoconductivity to degradation of the mechanical properties of the scaffold is rather challenging. Firstly, degradation rates *in vitro* are invariably different from those *in vivo*, and secondly, adding osteoconductive materials (e.g., TCP or HA) or osteoinductive agents (BMP-2) affects degradation. Here, osteoconductive refers to the ability of the scaffold to support the growth of capillaries and the cells that remodel bone (osteoclasts and osteoblasts); whereas, osteoinduction is the ability of the scaffold to recruit progenitor cells and then cause them to differentiate into osteoblasts. Thus, advances in tissue engineering come from both biology as well as material science.

5.7   DETERMINANTS OF DEGRADATION

TCP degrades 3–12 times faster than that of HA [63], and so mixtures of the two at different concentrations can vary degradation rates. Degradation control is ultimately dictated by the degrading mechanism. For the bioactive ceramics, degradation occurs by dissolution in which the solubility of the material affects the degradation rate. The pH of solution and ionic concentrations (e.g., substituting Mg^{2+} into HA) also affect the degradation rate of ceramics. Since dissolution basically involves the ceramic dissolving in water, the surface area of the scaffold to solution volume ratio affects the degradation rate. That is, an increase in scaffold porosity causes an increase in degradation, but an increase in crystallinity causes a decrease. Lastly, regions of high stress tend to degrade faster than regions of low stress within the same scaffold.

Polymer-based scaffolds typically degrade by hydrolysis, but enzymatic attack is a possibility. There are chemical groups in the polymer that are susceptible to cleavage in the presence of water to varying degrees, so degradation can be controlled by the choices of the structural units of the carbon-carbon backbone. As with bioactive ceramics, degradation rate depends on the access of water to these sites that have the chemical groups susceptible to hydrolysis. Thus, increasing crystallinity, decreasing material surface to volume ratio, and increasing inter-chain bonding all effectively reduce the degradation rate of the polymer.

5.8   CONCLUSION

Given the complexity of balancing degradation and tissue regeneration, a number of fabrication techniques have been attempted to achieve the right initial mechanical properties, degradation characteristics, and rate of osteogenesis such as salt leaching, solid form fabrication, nano-fiber spinning, and 3D printing, to name a few [71]. These techniques attempt to maintain mechanical strength of the porous scaffold during the process of bone regeneration. In regions of trabecular bone, the strength requirements are such that composite scaffolds having the requisite porosity for osteogenesis are achievable. When cortical bone defects need repair, porosity is currently a liability. Nonetheless, several strategies are being explored to achieve the mechanical demands of cortical bone while still regenerating the bone defect. Besides fabrication techniques, they include (1) generating tissue *in vitro*, or in some instances *in vivo* [77], before implantation and (2) the use of low porosity composites that provide sufficient bioresorption and release of growth factors to support osteogenesis.

REFERENCES

[1] Praemer, A, Furner S, Ricc DP. Musculoskeletal conditions in the united states. American Academy of Orthopaedic Surgeons; 1999. 5.1

[2] Goldberg, VM. Contemporary total joint arthroplasty. N C Med J 2007;68:447–50. 5.1

[3] Temenoff, JS, Mikos AG. Biomaterials : The intersection of biology and materials science. Pearson/Prentice Hall; 2008. 5.2.1, 5.2.1, 5.2.1, 5.3.1, 5.4.1

[4] Einhorn, TA, Simon SR. Orthopaedic basic science : Biology and biomechanics of the musculoskeletal system. American Academy of Orthopaedic Surgeons; 2000, p. 182–215. 5.1, 5.2.1, 5.1, 5.3.1

[5] Park, JB, Lakes RS. Biomaterials : An introduction. Plenum Press; 1992. 5.2.1, 5.2.3, 5.3.2

[6] Lewis, G, Carroll M. Effect of crosslinking UHMWPE on its tensile and compressive creep performance. Biomed Mater Eng 2001;11:167–83. 5.2.1

[7] Muggli, DS, Burkoth AK, Anseth KS. Crosslinked polyanhydrides for use in orthopedic applications: Degradation behavior and mechanics. J Biomed Mater Res 1999;46:271–8. DOI: 10.1002/(SICI)1097-4636(199908)46:2%3C271::AID-JBM17%3E3.0.CO;2-X 5.2.1

[8] Dean, D, Topham NS, Meneghetti SC, Wolfe MS, Jepsen K, He S, et al. Poly(propylene fumarate) and poly(dl-lactic-co-glycolic acid) as scaffold materials for solid and foam-coated composite tissue-engineered constructs for cranial reconstruction. Tissue Eng 2003;9:495–504. 5.2.2

[9] Ibim, SM, Uhrich KE, Bronson R, El-Amin SF, Langer RS, Laurencin CT. Poly(anhydride-co-imides): In vivo biocompatibility in a rat model. Biomaterials 1998;19:941–51. 5.2.2

[10] Holy, CE, Dang SM, Davies JE, Shoichet MS. In vitro degradation of a novel poly(lactide-co-glycolide) 75/25 foam. Biomaterials 1999;20:1177–85. DOI: 10.1016/S0142-9612(98)00256-7 5.2.2, 5.5

[11] Holy, CE, Cheng C, Davies JE, Shoichet MS. Optimizing the sterilization of plga scaffolds for use in tissue engineering. Biomaterials 2001;22:25–31. DOI: 10.1016/S0142-9612(00)00136-8 5.2.2

[12] Guelcher, S, Srinivasan A, Hafeman A, Gallagher K, Doctor J, Khetan S, et al. Synthesis, in vitro degradation, and mechanical properties of two-component poly(ester urethane)urea scaffolds: Effects of water and polyol composition. Tissue Eng 2007;13:2321–33. 5.2.2

[13] Dickson, G, Buchanan F, Marsh D, Harkin-Jones E, Little U, McCaigue M. Orthopaedic tissue engineering and bone regeneration. Technol Health Care 2007;15:57–67. 5.2.2, 5.5

[14] Guelcher, SA, Hollinger JO, An introduction to biomaterials. CRC/Taylor & Francis; 2006. 5.2.2

[15] Athanasiou, KA, Agrawal CM, Barber FA, Burkhart SS. Orthopaedic applications for pla-pga biodegradable polymers. Arthroscopy 1998;14:726–37. DOI: 10.1016/S0749-8063(98)70099-4 5.2.2

[16] Kanczler, JM, Oreffo RO. Osteogenesis and angiogenesis: The potential for engineering bone. Eur Cell Mater 2008;15:100–14. 5.2.2

[17] Tortelli, F, Cancedda R. Three-dimensional cultures of osteogenic and chondrogenic cells: A tissue engineering approach to mimic bone and cartilage in vitro. Eur Cell Mater 2009;17:1–14. 5.2.2

[18] Puelacher, WC, Vacanti JP, Ferraro NF, Schloo B, Vacanti CA. Femoral shaft reconstruction using tissue-engineered growth of bone. Int J Oral Maxillofac Surg 1996;25:223–8. DOI: 10.1016/S0901-5027(96)80035-X 5.2.2

[19] Ge, Z, Tian X, Heng BC, Fan V, Yeo JF, Cao T. Histological evaluation of osteogenesis of 3d-printed poly-lactic-co-glycolic acid (plga) scaffolds in a rabbit model. Biomed Mater 2009;4:21001. DOI: 10.1088/1748-6041/4/2/021001 5.2.2

[20] Liu, X, Ma PX. Polymeric scaffolds for bone tissue engineering. Ann Biomed Eng 2004;32:477–86. DOI: 10.1023/B:ABME.0000017544.36001.8e 5.2.2

[21] Lutolf, MP, Weber FE, Schmoekel HG, Schense JC, Kohler T, Muller R, et al. Repair of bone defects using synthetic mimetics of collagenous extracellular matrices. Nat Biotechnol 2003;21:513–8. DOI: 10.1038/nbt818 5.2.2

[22] Miyamoto, S, Takaoka K. Bone induction and bone repair by composites of bone morphogenetic protein and biodegradable synthetic polymers. Ann Chir Gynaecol Suppl 1993;207:69–75. 5.2.2

[23] Burdick, JA, Anseth KS. Photoencapsulation of osteoblasts in injectable rgd-modified peg hydrogels for bone tissue engineering. Biomaterials 2002;23:4315–23. DOI: 10.1016/S0142-9612(02)00176-X 5.2.2

[24] Guelcher, SA. Biodegradable polyurethanes: Synthesis and applications in regenerative medicine. Tissue Eng Part B Rev 2008;14:3–17. DOI: 10.1089/teb.2007.0133 5.2.2, 5.2.2

[25] Guelcher, SA, Gallagher KM, Didier JE, Klinedinst DB, Doctor JS, Goldstein AS, et al. Synthesis of biocompatible segmented polyurethanes from aliphatic diisocyanates and diurea diol chain extenders. Acta Biomater 2005;1:471–84. DOI: 10.1016/j.actbio.2005.02.007 5.2.2

[26] Kavlock, KD, Pechar TW, Hollinger JO, Guelcher SA, Goldstein AS. Synthesis and characterization of segmented poly(esterurethane urea) elastomers for bone tissue engineering. Acta Biomater 2007;3:475–84. DOI: 10.1016/j.actbio.2007.02.001 5.2.2

[27] Gogolewski, S, Gorna K, Turner AS. Regeneration of bicortical defects in the iliac crest of estrogen-deficient sheep, using new biodegradable polyurethane bone graft substitutes. J Biomed Mater Res A 2006;77:802–10. 5.2.2

[28] Gogolewski, S, Gorna K. Biodegradable polyurethane cancellous bone graft substitutes in the treatment of iliac crest defects. J Biomed Mater Res A 2007;80:94–101. 5.2.2

[29] Tierney, CM, Haugh MG, Liedl J, Mulcahy F, Hayes B, O'Brien FJ. The effects of collagen concentration and crosslink density on the biological, structural and mechanical properties of collagen-gag scaffolds for bone tissue engineering. J Mech Behav Biomed Mater 2009;2:202–9. DOI: 10.1016/j.jmbbm.2008.08.007 5.2.3

[30] Bedran-Russo, AK, Pashley DH, Agee K, Drummond JL, Miescke KJ. Changes in stiffness of demineralized dentin following application of collagen crosslinkers. J Biomed Mater Res B Appl Biomater 2008;86B:330–4. DOI: 10.1002/jbm.b.31022 5.2.3

[31] Lin, H, Zhao Y, Sun W, Chen B, Zhang J, Zhao W, et al. The effect of crosslinking heparin to demineralized bone matrix on mechanical strength and specific binding to human bone morphogenetic protein-2. Biomaterials 2008;29:1189–97. DOI: 10.1016/j.biomaterials.2007.11.032 5.2.3

[32] Catanese, J, 3rd, Iverson EP, Ng RK, Keaveny TM. Heterogeneity of the mechanical properties of demineralized bone. J Biomech 1999;32:1365–9. DOI: 10.1016/S0021-9290(99)00128-1 5.2.3

[33] Bowman, SM, Zeind J, Gibson LJ, Hayes WC, McMahon TA. The tensile behavior of demineralized bovine cortical bone. J Biomech 1996;29:1497–501. DOI: 10.1016/0021-9290(96)84546-5 5.2.3

[34] Viguet-Carrin, S, Garnero P, Delmas PD. The role of collagen in bone strength. Osteoporos Int 2006;17:319–36. DOI: 10.1007/s00198-005-2035-9 5.2.3

[35] Wollensak, G, Iomdina E. Long-term biomechanical properties after collagen crosslinking of sclera using glyceraldehyde. Acta Ophthalmol 2008;86:887–93. DOI: 10.1111/j.1755-3768.2007.01156.x 5.2.3

[36] Tuli, SM, Singh AD. The osteoinductive property of decalcified bone matrix. An experimental study. J Bone Joint Surg Br 1978;60:116–23. 5.2.4

[37] Rabie, AB, Wong RW, Hagg U. Composite autogenous bone and demineralized bone matrices used to repair defects in the parietal bone of rabbits. Br J Oral Maxillofac Surg 2000;38:565–70. DOI: 10.1054/bjom.2000.0464 5.2.4

[38] Veillette, CJ, McKee MD. Growth factors–bmps, dbms, and buffy coat products: Are there any proven differences amongst them? Injury 2007;38 Suppl 1:S38–48. DOI: 10.1016/j.injury.2007.02.009 5.2.4

[39] Werntz, JR, Lane JM, Burstein AH, Justin R, Klein R, Tomin E. Qualitative and quantitative analysis of orthotopic bone regeneration by marrow. J Orthop Res 1996;14:85–93. DOI: 10.1002/jor.1100140115 5.2.4

[40] Qiu, QQ, Connor J. Effects of gamma-irradiation, storage and hydration on osteoinductivity of dbm and dbm/am composite. J Biomed Mater Res A 2008;87:373–9. DOI: 10.1002/jbm.a.31790 5.2.4

[41] Guelcher, SA, Hollinger JO, An introduction to biomaterials. CRC/Taylor & Francis; 2006, p. 249–59. 5.2.4

[42] Di, Martino A, Sittinger M, Risbud MV. Chitosan: A versatile biopolymer for orthopaedic tissue-engineering. Biomaterials 2005;26:5983–90. DOI: 10.1016/j.biomaterials.2005.03.016 5.2.4

[43] Park, YJ, Lee YM, Park SN, Sheen SY, Chung CP, Lee SJ. Platelet derived growth factor releasing chitosan sponge for periodontal bone regeneration. Biomaterials 2000;21:153–9. DOI: 10.1016/S0142-9612(99)00143-X 5.2.4

[44] Seol, YJ, Lee JY, Park YJ, Lee YM, Young K, Rhyu IC, et al. Chitosan sponges as tissue engineering scaffolds for bone formation. Biotechnol Lett 2004;26:1037–41. DOI: 10.1023/B:BILE.0000032962.79531.fd 5.2.4

[45] Zhang, Y, Zhang M. Synthesis and characterization of macroporous chitosan/calcium phosphate composite scaffolds for tissue engineering. J Biomed Mater Res 2001;55:304–12. DOI: 10.1002/1097-4636(20010605)55:3%3C304::AID-JBM1018%3E3.0.CO;2-J 5.2.4

[46] Habibovic, P, de Groot K. Osteoinductive biomaterials–properties and relevance in bone repair. J Tissue Eng Regen Med 2007;1:25–32. DOI: 10.1002/term.5 5.3.2

[47] El-Ghannam, A. Bone reconstruction: From bioceramics to tissue engineering. Expert Rev Med Devices 2005;2:87–101. DOI: 10.1586/17434440.2.1.87 5.3.2

[48] Guelcher, SA, Hollinger JO, An introduction to biomaterials. CRC/Taylor & Francis; 2006, p. 311–39. 5.3.2

[49] Ramay, HR, Zhang M. Biphasic calcium phosphate nanocomposite porous scaffolds for load-bearing bone tissue engineering. Biomaterials 2004;25:5171–80. DOI: 10.1016/j.biomaterials.2003.12.023 5.3.2

[50] Ramay, HR, Zhang M. Preparation of porous hydroxyapatite scaffolds by combination of the gel-casting and polymer sponge methods. Biomaterials 2003;24:3293–302. 5.3.2

[51] Ogino, M, Ohuchi F, Hench LL. Compositional dependence of the formation of calcium phosphate films on bioglass. J Biomed Mater Res 1980;14:55–64. DOI: 10.1002/jbm.820140107 5.3.2

[52] Schepers, E, de Clercq M, Ducheyne P, Kempeneers R. Bioactive glass particulate material as a filler for bone lesions. J Oral Rehabil 1991;18:439–52. 5.3.2

[53] Oonishi, H, Kushitani S, Yasukawa E, Iwaki H, Hench LL, Wilson J, et al. Particulate bioglass compared with hydroxyapatite as a bone graft substitute. Clin Orthop Relat Res 1997:316–25. 5.3.2

[54] Klein, CP, Driessen AA, de Groot K, van den Hooff A. Biodegradation behavior of various calcium phosphate materials in bone tissue. J Biomed Mater Res 1983;17:769–84. DOI: 10.1002/jbm.820170505 5.3.2

[55] Liao, S, Wang W, Uo M, Ohkawa S, Akasaka T, Tamura K, et al. A three-layered nanocarbonated hydroxyapatite/collagen/plga composite membrane for guided tissue regeneration. Biomaterials 2005;26:7564–71. DOI: 10.1016/j.biomaterials.2005.05.050 5.3.3

[56] Li, Z, Yubao L, Aiping Y, Xuelin P, Xuejiang W, Xiang Z. Preparation and in vitro investigation of chitosan/nano-hydroxyapatite composite used as bone substitute materials. J Mater Sci Mater Med 2005;16:213–9. DOI: 10.1007/s10856-005-6682-3 5.3.3

[57] McManus, AJ, Doremus RH, Siegel RW, Bizios R. Evaluation of cytocompatibility and bending modulus of nanoceramic/polymer composites. J Biomed Mater Res A 2005;72:98–106. DOI: 10.1002/jbm.a.30204 5.3.3

[58] Kaito, T, Myoui A, Takaoka K, Saito N, Nishikawa M, Tamai N, et al. Potentiation of the activity of bone morphogenetic protein-2 in bone regeneration by a pla-peg/hydroxyapatite composite. Biomaterials 2005;26:73–9. DOI: 10.1016/j.biomaterials.2004.02.010 5.3.3

[59] Tancred, DC, Carr AJ, McCormack BA. The sintering and mechanical behavior of hydroxyapatite with bioglass additions. J Mater Sci Mater Med 2001;12:81–93. DOI: 10.1023/A:1026773522934 5.3.3

[60] Liao, SS, Cui FZ, Zhang W, Feng QL. Hierarchically biomimetic bone scaffold materials: Nano-ha/collagen/pla composite. J Biomed Mater Res B Appl Biomater 2004;69:158–65. 5.3.3

[61] Burg, KJ, Porter S, Kellam JF. Biomaterial developments for bone tissue engineering. Biomaterials 2000;21:2347–59. DOI: 10.1016/S0142-9612(00)00102-2 5.3.3, 5.5

[62] Boccaccini, AR, Blaker JJ. Bioactive composite materials for tissue engineering scaffolds. Expert Rev Med Devices 2005;2:303–17. DOI: 10.1586/17434440.2.3.303 5.3.3

[63] Hutmacher, DW, Schantz JT, Lam CX, Tan KC, Lim TC. State of the art and future directions of scaffold-based bone engineering from a biomaterials perspective. J Tissue Eng Regen Med 2007;1:245–60. DOI: 10.1002/term.24 5.3.3, 5.5, 5.7

[64] Hasegawa, S, Ishii S, Tamura J, Furukawa T, Neo M, Matsusue Y, et al. A 5–7 year in vivo study of high-strength hydroxyapatite/poly(l-lactide) composite rods for the internal fixation of bone fractures. Biomaterials 2006;27:1327–32. 5.3.3

[65] Shen, L, Yang H, Ying J, Qiao F, Peng M. Preparation and mechanical properties of carbon fiber reinforced hydroxyapatite/polylactide biocomposites. J Mater Sci Mater Med 2009. DOI: 10.1007/s10856-009-3785-2 5.3.3

[66] Li, JP, de Wijn JR, van Blitterswijk CA, de Groot K. Porous ti6al4v scaffolds directly fabricated by 3d fibre deposition technique: Effect of nozzle diameter. J Mater Sci Mater Med 2005;16:1159–63. DOI: 10.1007/s10856-005-4723-6 5.4.1

[67] Li, JP, Habibovic P, van den Doel M, Wilson CE, de Wijn JR, van Blitterswijk CA, et al. Bone ingrowth in porous titanium implants produced by 3d fiber deposition. Biomaterials 2007;28:2810–20. DOI: 10.1016/j.biomaterials.2007.02.020 5.4.2

[68] Chen, F, Feng X, Wu W, Ouyang H, Gao Z, Cheng X, et al. Segmental bone tissue engineering by seeding osteoblast precursor cells into titanium mesh-coral composite scaffolds. Int J Oral Maxillofac Surg 2007;36:822–7. DOI: 10.1016/j.ijom.2007.06.019 5.4.2

[69] Vehof, JW, Haus MT, de Ruijter AE, Spauwen PH, Jansen JA. Bone formation in transforming growth factor beta-i-loaded titanium fiber mesh implants. Clin Oral Implants Res 2002;13:94–102. DOI: 10.1034/j.1600-0501.2002.130112.x 5.4.2

[70] Vehof, JW, Mahmood J, Takita H, van't Hof MA, Kuboki Y, Spauwen PH, et al. Ectopic bone formation in titanium mesh loaded with bone morphogenetic protein and coated with calcium phosphate. Plast Reconstr Surg 2001;108:434–43. 5.4.2

[71] Ge, Z, Jin Z, Cao T. Manufacture of degradable polymeric scaffolds for bone regeneration. Biomed Mater 2008;3:22001. 5.5, 5.8

[72] Karageorgiou, V, Kaplan D. Porosity of 3d biomaterial scaffolds and osteogenesis. Biomaterials 2005;26:5474–91. DOI: 10.1016/j.biomaterials.2005.02.002 5.5

[73] Burg, KJ, Holder WD, Culberson CR, Beiler RJ, Greene KG, Loebsack AB, et al. Parameters affecting cellular adhesion to polylactide films. J Biomater Sci Polym Ed 1999;10:147–61. DOI: 10.1163/156856299X00108 5.5

[74] Hutmacher, DW. Scaffolds in tissue engineering bone and cartilage. Biomaterials 2000;21:2529–43. DOI: 10.1016/S0142-9612(00)00121-6 5.5

[75] van, den Dolder J, Bancroft GN, Sikavitsas VI, Spauwen PH, Jansen JA, Mikos AG. Flow perfusion culture of marrow stromal osteoblasts in titanium fiber mesh. J Biomed Mater Res A 2003;64:235–41. DOI: 10.1002/jbm.a.10365 5.5

[76] Martin, I, Wendt D, Heberer M. The role of bioreactors in tissue engineering. Trends Biotechnol 2004;22:80–6. DOI: 10.1016/j.tibtech.2003.12.001 5.5

[77] Holt, GE, Halpern JL, Dovan TT, Hamming D, Schwartz HS. Evolution of an in vivo bioreactor. J Orthop Res 2005;23:916–23. DOI: 10.1016/j.orthres.2004.10.005 5.8

[78] Zhang, R, Ma PX. Poly (alpha-hydroxyl acids)/hydroxyapatite porous composites for bone-tissue engineering. I. Preparation and morphology. J Biomed Mater Res 1999;44:446–55. 5.3.3

CHAPTER 6

Mechanical and Structural Evaluation of Repair/Tissue Engineered Bone

CHAPTER SUMMARY

Chapter 6 presents the information regarding the current methodologies for the mechanical and structural evaluation of repair and tissue engineered bone tissues. Four major categories are covered, which include traditional mechanical testing tests (e.g., tension/compression, torsion, bending, and indentation), *in vivo* stiffness measuring systems, non invasive imaging techniques, and numerical simulations.

6.1 INTRODUCTION

Mechanical and structural evaluation of repair/tissue engineered bone is an important aspect in both tissue engineering research and clinical applications. Test standards need to be established to evaluate the repaired and replacements tissues after surgery and determine whether these treatments are successful in pre-clinical animal models and clinical applications.

In vitro mechanical testing of the fracture callus of bone from animal models provides direct information regarding the recovery of biomechanical properties of repair tissue during healing process. Healing bone provides loading bearing capacity for the skeletal system. Therefore, the success of a certain tissue engineering strategy for the treatment of bone defects should be evaluated by the restoration of stiffness and strength of healing bone. A number of well-established mechanical testing methods such as torsional tests, bending tests, tensile tests, compressive tests, and nanoindentation tests have been introduced to measure biomechanical properties of fracture callus.

In clinical evaluation of bone healing as well as preclinical animal models, non-invasive advanced imaging techniques offer qualitative and quantitative information regarding the architectural, mineralization and structural stability of fracture callus during the healing process. Among these imaging techniques, plain x-ray radiography, dual energy x-ray absorptiometry (DEXA), quantitative computed tomography (QCT), and micro computed tomography (Micro-CT) have been commonly used in the evaluation of fracture healing.

In addition, in vivo stiffness measurement and finite element modeling of healing bone are available in clinical settings to monitor the biomechanical progressing of fracture healing. *In vivo* stiffness monitoring of fracture healing with external fixation allows the biomechanics of the fracture treatment to be fine tuned to provide the optimal environment for fracture healing with a tissue engineering strategy. Finite element models combined with computed tomography images can be used to calculate the material properties of fracture callus and provide biomechanical properties of healing bone.

This chapter is intended to describe some important aspects of the mechanical and structural evaluation of repair/tissue engineered bones, which include *in vitro*, *in vivo*, non-invasive evaluation methodologies as well as computational prediction of the healing process.

6.2   IN VITRO BIOMECHANICAL EVALUATION OF FRACTURE REPAIR

It is well established that biomechanical testing is essential for characterizing the effect of various scaffolds or growth factors on the healing processing in bone [1]. Most of the biomechanical tests, such as torsion and bending, in evaluating tissue engineered bone from in vivo animal models are based on structural testing. Due to irregular geometry in fracture callus, biomechanical testing has its limitations in providing useful biomechanical information about the intrinsic material properties of callus tissue. Furthermore, large variation in experimental data is expected due to significant differences in shape and material quality. Nevertheless, biochemical testing remains a quantitative way to assess the restoration of functional capacity for healing fractures.

The comparison of different studies on bone defects in large animal models is rather difficult because the defects produced had different sizes in different animal models. A standard technique for biomechanical testing of bone defects needs to be established to facilitate the comparison. A common practice is to measure the biomechanical properties of fractured tibia and the contralateral tibia without fracture. A percentage of stiffness or strength is used for comparison among different studies.

The biomechanical data on fracture healing is confounded by differences in testing methodologies in terms of mode of testing. Caution is needed when interpreting the biomechanical results of fracture healing due to differences in torsional, bending, tensile, and compressive testing.

6.2.1   TORSIONAL TEST

Torsional testing is the most widely used method in the biomechanical evaluation of fracture healing in non-critical size defects with growth factors as well as repairing segmental bone defects using tissue engineered bone. From the perspective of mechanics, the stress distribution of a callus during a torsion test is not uniform with the highest shear stress at the boundary and the lowest shear stress in the central axis. Therefore, torsional testing is a structural test and cannot provide material properties of fracture callus. Peak torsional strength, torsional stiffness, and the energy to failure of fracture calluses have been extensively reported in the literature.

Sample preparation for a torsional test involves fixation of the proximal and distal ends of the bone with a central fracture callus in a resin or embedding medium. Examples are PMMA [2–5], superglue [6], high performance polymers [7], polyester resin [8], epoxy [9], acrylate [8], and woods metal [10]. Proper axial alignment of the embedded bone sample is essential for avoiding bending of the sample during gripping and testing.

Torsion testing requires either a biaxial mechanical testing machine or a custom-designed torsion device. The proper loading rate is essential in torsion tests to replicate or simulate fractures in clinical situations in which high speed fractures occur such as ski fractures of the tibia and high energy trauma fractures in automobile accidents [11]. A special fixture for torsional tests (Figure 6.1) has been developed to allow a reproducible and rapid loading rate [12].

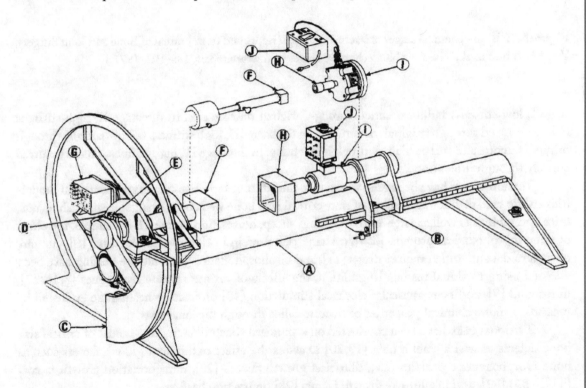

Figure 6.1: A custom-made torsional testing device to measure torsional properties of fracture callus. A. Frame; B. Tail stocks; C. Pendulum; D. Dog clutch in pendulum; E. Dog clutch on rotating shaft; F. Rotating grips; G. Angular deformation transducer; H. Stationary grips; I. Torque transducers; J. Self calibration system. (Reprinted from Journal of Biomechanics, Vol. 4, Burstein and Frankel, a standard test for laboratory animal bone, 155–158, 1971.)

Using this device, torsional tests were conducted on healing tibial fractures of non-critical size in rabbits. Four biomechanical stages of fracture healing (Figure 6.2) were defined [13]. In the

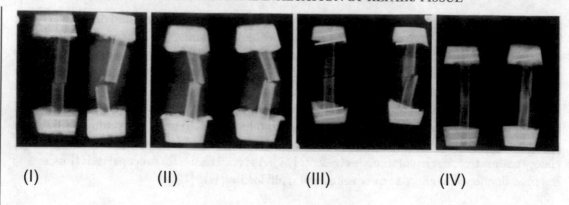

(I)   (II)   (III)   (IV)

Figure 6.2: Biomechanical stages of fracture healing. (Reprinted from Journal of Bone and Joint Surgery, Vol. 59, White et al., The four biomechanical stages of fracture repair, 188-192, 1977.)

stage I, low stiffness failure occurred through original fracture site. In the stage II, high stiffness failure occurred through original fracture site. In the stage III, high stiffness failure partially through original fracture site and partially through intact bone. In the stage IV, high stiffness failure occurred entirely through intact bone.

Torsional tests have been used to evaluate the effect of bioactive agents and treatment modalities in the case of fracture healing of non-critical size bone defects in small animals such as rats, mice and rabbits as well as large animals such as sheep, horses, and canines. For example, the effect of bone morphological proteins [3, 5], cysteine-rich protein [14], thrombin peptide [10], vitamin, non-steroidal anti-inflammatory drugs [15], and bisphosphonate [9] on fracture healing have been assessed using torsional testing. In addition, the effects of various fixation techniques [8, 16, 17], ultrasound [7], and neuromuscular electrical stimulation [18] on fracture healing are evaluated by measuring biomechanical properties of fracture callus through torsional tests.

Torsional tests have been conducted on a standard sheep tibia fracture model of critical size bone defects as well as rabbit tibia [19, 20] to assess the effect of different growth factors such as bone morphogenetic proteins [21], fibroblast growth factors [20], transformation growth factors (TGF-β3) [22], and insulin-like growth factors [23] on fracture healing.

Although torsional stiffness and torsional strength were reported in most studies, information regarding the pattern of fracture site from the destructive torsional test was rarely revealed. This information is essential in judging the stage of bone repair. In a study regarding the effect of bone morphologic proteins on the fracture healing of rabbit ulna osteomy model, fracture pattern of healing bone has been reported [5]. Some of the healing bones from BMP treated animals failed through the interface between intact bone and the osteotomy whereas the controls without BMP treatment failed through the original osteotomy sites. In another study of repairing segmental defect with a coral composite and BMP, it was reported that all explanted specimens failed with a consistent pattern of spiral fractures [21]. As discussed in a previous study [13], biomechanical stage of fracture healing can be defined as four stages. This study was conducted on the animal model of non-critical size bone defects. However, the biomechanical stage of fracture repair for critical-size bone defects is unknown at this time. The stiffness and strength of tissue engineering constructs may affect the biomechanical stages of fracture healing.

In the early stage of fracture healing, the interface between fracture callus and intact bone is relatively weak. A torsional test can break the interface between intact bone and fracture callus. Therefore, a failure through the original fracture is observed. Once the interface between intact bone and fracture callus is strong, the failure does not occur at the interface. Since the size of fracture callus is much larger than the intact bone, the fracture callus is not going to break during torsion. The weakest spot at this stage is the intact bone due to its small size. As a result, the torque applied on the bone will cause the fracture to occur through the intact bone. This emphasizes that torsional testing is a structural test, not a material test. To accurately estimate the material properties of fracture callus, other mechanical testing such as nanoindentation and nanoscratch test may be used.

6.2.2 BENDING

Bending tests are often used in assessing fracture healing since they are convenient and efficient. Sample preparation for a bending test does not involve embedding the bone into a fixation medium. Bending test can be performed in either a three-point or a four-point configuration. Bending involves complex stress fields since the fracture callus is subject to a complex pattern of tension, compression and possible shear, which complicates the interpretation of data. As a result, bending testing is considered as a structural test. The structural properties can be converted to the material properties by taking account of specimen geometry if beam theory is assumed. However, the underlying assumptions are difficult to justify due to irregular geometry of the fracture callus. Therefore, modulus of the fracture callus is rarely reported in the bending test. In the case of a non-destructive test, bending stiffness of fracture callus is often reported in the literature. In the case of a destructive test, both bending stiffness and strength are recorded.

Destructive four-point bending test has been used on rat tibial models with a segmental bone defect size of 4 mm to test the effect of basic fibroblast growth factor on fracture healing [24]. Bending strength, stiffness and energy to failure of tibial specimens were reported for this study. Destructive three-point bending test has been conducted on rabbit tibial models with segmental bone defects of 5 mm in length to study the effects of a complex of β-tricalcium phosphate granules, collagen, and fibroblast growth factor-2 on fracture healing [20]. Bending strength and bending stiffness have been reported for this study. Non-destructive four-point bending test has been performed on sheep tibia with a defect size of 25 mm to test the effect of a platelet-rich plasma on fracture healing [25]. Whole tibiae were removed after the completion of experiments. The free length of tibiae during the test was 240 mm while the loaded region had a length of 5 mm with the fracture callus in the center. The flexural rigidity was reported for this study.

6.2.3 TENSION

Tension tests provide information on the elastic properties of fracture callus since uniaxial tension provides a uniform stress field to the sample. In this sense, tensile testing is a material test, not a structural test. The material properties of fracture callus can be calculated after accounting for its geometry. A few studies have used tensile tests to measure biomechanical properties of fracture callus [26–34]. Tensile strength measurements of fractures are more useful than torsional and bending values during the first phases of fracture repair because they constitute a sensitive method of fracture repair [31,35].

Tensile tests of whole fracture callus involves potting two ends of whole bone (e.g., femur or tibia) into a tube with embedding medium, such as PMMA and other plastic resins [27,29]. The prepared samples are then tested under a mechanical testing system with a custom-designed fixture. Tensile strength, stiffness, and energy to failure are normally reported in the literature.

Tensile tests of fractional fracture callus can be used to examine the effect of size and stability of gap on the fracture healing using animal models, such sheep metatarsal osteotomy. In a previous study, for example, sheep bone specimens of $2\times2\times60$ mm were taken from dorsal and plantar sections of fracture callus at metatarsal, and then tensile testing was performed on such specimens until failure at 5 mm/min [26]. The tensile strength of callus was used to assess the effect of fracture size and stability on the healing.

6.2.4 COMPRESSION

Only a few studies have used compression testing to assess the biomechanical properties of fracture callus [27,36]. In one study, compression tests were performed on a 3.6 mm slice sectioned from the thickest point of the fracture callus [27]. Bone samples were compressed against two parallel platens. A tilting platen was used to minimize alignment error. Compressive strength was calculated from peak load and cross-sectional area from QCT measurements of fracture callus. Similarly, compressive strength of fracture callus was measured in a goat tibia model with a segmental defect of 2.6 cm in length by trimming the central part of fracture callus [36].

6.2.5 MICRO/NANO INDENTATION TESTS

Indentation tests are a promising tool to assess the local tissue properties of regenerated bone tissues (e.g., fracture callus) [26, 37–42]. Both microindentation and nanoindentation tests have been used for this purpose. Microindentation tests have been commonly employed in studies of *in vivo* tissue regeneration of large animal models. For instance, the flat-surfaced and spherical-surfaced indenters (1.5 mm in diameter) have been successfully employed to measure the material properties of fracture callus in osteotomy models of canine tibia [38], sheep tibia [39], and sheep metatarsal [26]. The indentation stiffness is a major tissue property obtained in the indentation tests, which can be determined by calculating the slope of the linear region of the load-deformation curve during indentation tests [37].

Recent advances in nanoindentation technologies have allowed for *in situ* measurements of material properties of regenerated tissues. Nanoindentation tests are especially suitable for studies using small animal models, such as rats and mice [41–43], in which the size limitation of repairing bone samples is critical for the success of experiments. For example, nanoindentation tests were performed on a 200μm thick section from the fracture callus of rat femur with the closed fracture [42]. The elastic modulus of fracture callus was calculated from the loading-displacement curve using the Oliver–Pharr method [44]. The elastic modulus of regenerated tissues by nanoindentation tests has been shown to correlate with tissue mineral density determined by micro-CT measurements [43]. In addition, nanoindentation tests can also be used for large animal models. In this case, the spatial and temporal variations of mechanical properties of fracture callus become the focus of research [40].

6.3 IN VIVO STIFFNESS MONITORING OF HEALING BONE

In vivo evaluation of mechanical and structural property of the tissue engineered and repair bone tissue is rather a challenging endeavor. Nonetheless, bioengineers and researchers have made a good effort in this regard. For example, the evolution of callus stiffness during fracture healing can be predicted when a strain gage is cemented on the beam of an external fixation. As a result, *in vivo* measurements of bending stiffness become possible. In this case, numerical simulations can be performed to analyze the relationship between the bending rigidity of fracture callus and the percentage of fracture healing [45]. Based on the assumptions of beam theory, the relationship between deformation and stiffness of callus could be finally estimated.

In addition, direct measurements of *in vivo* bending stiffness have been performed to monitor the biomechanical progression of fracture healing with external fixations [46–48]. The general approach for *in vivo* stiffness measurement is to apply a load across the fracture gap and measure the resultant deflections (Figure 6.3). For example, a custom-made device has been made to serve the purpose. With force and displacement transducers being attached to the external pins, a displacement-controlled bending moment is created in a transverse axis by means of fastening screws. The force-displacement curve is recorded and the bending stiffness is calculated from the slope of the linear region of the curve. As the fracture heals, the increasing slope of the load-versus-displacement curve indicates a return to the original stiffness of the bone. This device has been applied to a standard sheep tibia model to test the influence of growth factors on bone regeneration in a segmental defect in the tibia [49]. There are two benefits for the *in vivo* stiffness monitoring. First, such measurements allow for the early action of treatment correction if delayed union is expected. Second, the mechanical stimulus can be fine-tuned during the fracture treatment to provide the optimal environment for fracture healing.

In fact, similar techniques have been used in clinical practice. For example, *in vivo* stiffness measurements have been performed on patients to assess bone healing after fracture or lengthening in adults [50,51] as well as children [52]. Patients treated by external fixator for tibia fractures are recruited for assessing their fracture healing with a device that can measure the fracture stiffness of the repair bone [50]. The device includes a system of electronic goniometer, force plate, and micro-

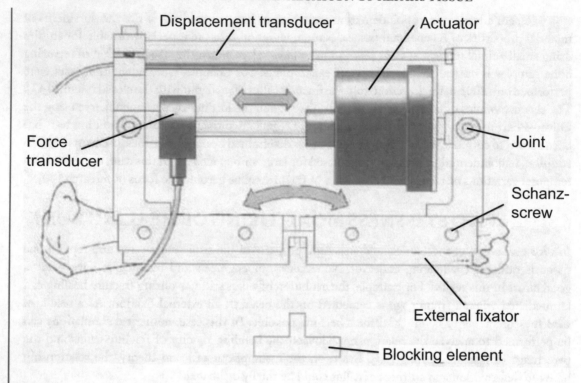

Figure 6.3: In vivo stiffness measurement device mounted onto an external fixator. The device contains an actuator, a piezoelectric force transducer, and a displacement transducer. Blocking element is removed from external fixator bar before stiffness measurement. Bending moment is produced by adjusting four Schanz-screws transverse to the tibia axis. (Reprinted from Journal of Orthopaedic Research, Vol. 21, 2003, Lill et al., Page 836-842.)

processor, which attaches to the pins of external fixation. The heel is supported on the force plate, enabling measurement of the force required to produce a measured angular deflection (Figure 6.4). By pressing on the fracture site and causing an angular deflection of the low limb in the sagittal plane, the stiffness of the fracture can be calculated. The device has been validated to provide a measure of fracture stiffness [51].

Although the stiffness of healing bones can be accurately measured with external fixators, caution must be taken to use such stiffness measurements as a criterion of healing since it does not necessarily reflect structural strength of the healing bone [53,54]. Some experimental studies on the biomechanical behavior of healing bones in canine radii have shown that the weight-bearing radii may regain bending stiffness more rapidly than bending strength [54]. A study using rabbit radius fracture models has also shown that bone strength recovers more slowly than bone stiffness during fracture healing [53].

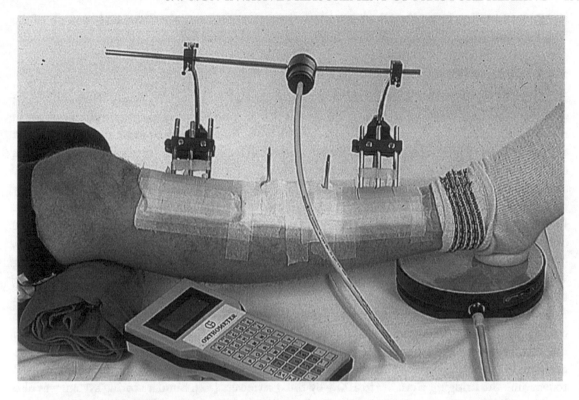

Figure 6.4: A photograph of the device used to measure the in vivo stiffness of a patient's tibia in clinical practice. (Reprinted from Clinical Biomechanics, Vol. 15, 2000, Eastaugh-Waring et al., Page 140-142.)

6.4   NON-INVASIVE MEASUREMENT OF FRACTURE HEALING

In addition to mechanical testing, advanced bioimaging techniques can be used to evaluate the structure of fracture callus. In clinical practice, X-ray images have been normally used to estimate and monitor bone formation and to determine the point of return to full load-bearing capacity and removal/degradation of orthopaedic implants. However, plain radiographs give only qualitative information about the degree of fracture healing. With the development of advanced bioimaging techniques in recent decades, quantitative methods such as plain radiography, dual energy x-ray absorpiometry (DEXA), quantitative computed tomography (QCT), micro computed tomography (micro-CT), and ultrasound techniques have been used to monitor the effect of tissue engineered bone on fracture healing. Although these quantitative methods provide information such as bone mineral density, area, and microstructure of fracture callus that contributes to the biomechanical integrity of healing bone, they don't provide a direct measurement of stiffness and strength of the healing callus [47].

6.4.1   X-RAY (PLAIN RADIOGRAPHY)

X-ray can be employed to assess the effect of various strategies on repairing bone defects at different time points. The advantage of X-ray technique is that it is simple, easy to operate and available in most hospitals. After surgery, X-ray images can be taken from patients or experimental animals at different stages of healing and at the different anatomical locations of the fracture site. X-ray images can be scanned using an optical scanner, thus allowing for quantifying callus area and intensity changes using image-processing techniques [21]. Based on the radio opacity, the relative density and distribution of mineralized bony tissues at different fracture healing stages can be determined. However, such measurements are not quantitative but at most qualitative for evaluating the degree of fracture healing. For example, X-ray was used to examine the healing progress of rabbit tibia at 2, 4 and 8 weeks after surgery [20] [22]. In most studies, it is used to visually compare fracture healing with different strategies [55]. To ensure consistency between the studies, a semi-quantitative scoring system has been developed to evaluate the effect of fracture healing [19]. According to the pattern from X-ray pictures, a score of 0 to 6 is assigned, in which zero (0) indicates non-unions of fracture and six (6) indicates the highest radio opacity in the defect [19]. Another score system has also been employed to examine the bridge side of callus [56]. In this scoring system, zero (0) means no bridge and four (4) means bridging at anterior, posterior, lateral, and medial sides [56]. This scale was used to examine the bridging from 0% to 100% by three reviewers [57]. Similar scoring system was also used by other investigators for evaluating healing of critical bone defects [25].

X-ray images are rarely used alone to evaluate the fracture healing due to its qualitative feature although such techniques are employed in many studies for fracture healing evaluation. Additional techniques, such as advanced image methods [20], histological methods, and biomechanical testing, are combined with X-ray images to assess the fracture healing in studies of the efficacy of bioactive agents or various scaffolds in tissue regeneration.

6.4.2   DUAL-ENERGY X-RAY ABSORPTIOMETRY (DEXA)

Dual-energy X-ray absorptiometric (DEXA) methods can provide a quantitative measure of bone mineral density and bone mineral content in healing bone [37, 58–63]. DEXA is able to account for variations in the amount of overlaying soft tissue by measuring the attenuations of X-ray beams at two distinct energies as beams pass through bone and soft tissue. The thicknesses of bone and soft tissue can be calculated by solving two equations representing the attenuation of x-ray beam intensities at different energies [64]. After taking into consideration of the thicknesses of bone and soft tissue, the confounding effect of soft tissue can be separated. The measurement of bone mineral content is done by projection of a three-dimensional region of fracture callus into a plane. Bone mineral density of the healing tissue is obtained by normalizing bone mineral content against the projected plane.

Measurement of bone mineral density by DEXA has been shown to be useful in estimating strength of fracture callus *in vitro*. The relationship between bone mineral density within a segmental defect of 30 mm measured by DEXA and biomechanical evaluation of the healing bone has been assessed in a standard sheep tibia model [60]. Results for this study have shown that bone mineral density is significantly correlated with torsional strength of fracture callus [60]. In addition, another study of healing canine tibial osteotomy of 2 mm has shown that the bone mineral density of fracture callus by DEXA has strong correlations with a signification relation between ultimate torque and torsional stiffness [37]. As a result, DEXA becomes a useful tool for noninvasively monitoring the effects of various tissue-engineering strategies on the treatment of fracture healings.

6.4.3 COMPUTED TOMOGRAPHY (CT) AND QUANTITATIVE COMPUTED TOMOGRAPHY (QCT)

Computed tomography (CT) is available in most hospitals and research facilities and allows in-site evaluation of bony formation within the tissue engineered bone implants in fracture healing. Therefore, it has been used in the pre-clinical animal models for developing tissue-engineering strategies to treat bone defects. Another advantage of CT is that it can be utilized to assess the cross-section area and mineralization density of fracture callus [21].

During fracture healing, newly formed bone within and/or around implants can be clearly visualized in CT images as shown in Figure 6.5 [65]. These CT images can be used to quantify the amount of bone formation in fracture callus. In addition, CT can be used to estimate biodegradation of tissue engineered constructs in segmental bone defects. For example, this technique has shown that some of the constructs fragment and some remain intact until the end of the experiment [21]. By taking consecutive CT scans with a layer thickness of 1 mm and reconstructing the slices in to a three-dimensional image, the densitometric moment of inertia of fracture callus can be calculated. The efficacy of the approach has been verified in a sheep tibia fracture model with an osteotomy gap of 3 mm [66].

It is worthy to mention that the density of the callus in computed tomography is expressed in Hounsfield units, which is different from areal bone mineral density measured by DEXA. Therefore, the measurement of density in callus is semi-quantitative in computed tomography. To quantify the bone mineral density of fracture callus, quantitative computed tomography (QCT) are needed.

Similar to DEXA, QCT is also a noninvasive and quantitative technique based on the measurement of X-ray attenuation. The difference is that DEXA measures the areal mineral density of bone, whereas QCT can measure the volumetric bone mineral density values. Quantitative evaluation of QCT comes from regular computed tomography used for patients in the hospital. A QCT scan of an anatomical site is acquired through regular computed tomography in conjunction with a calibration phantom that is made of series of hydroxyapatite standards with various concentrations. QCT has been extensively used as a tool for assessments of the volume and mineral density of fracture callus during fracture healing in both small animals [4,5,7,9,18] and large animals [25,46,49,58,67].

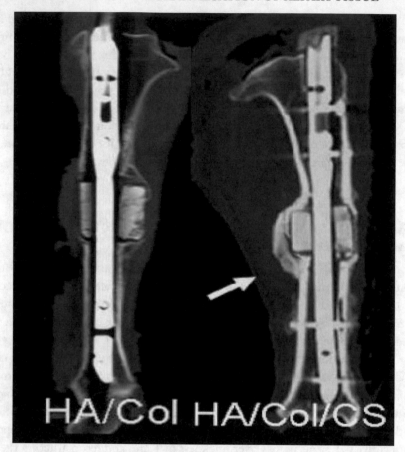

Figure 6.5: CT scans of hydroxyapatite/collagen and hydroxyapatite/collagen/ chondroitin sulphate implants after 3 month showing a considerable callus reaction (arrows) around HA/Col/CS implants only. (Reprinted from Schneiderss paper published in Journal of Orthopaedic Research, 2009, Vol. 27, Page 15-21.)

QCT scans of healing bone can be performed *in vivo* to quantify changes of volumetric bone mineral density of fracture callus at different stages of fracture healing under general anesthesia [7, 18, 46, 49]. In experimental studies, post-mortem QCT scans of healing bone could also be done after it is removed from animals [4, 5, 9, 25, 58, 67]. During the *in vitro* scanning of healing bone, it is immersed in alcohol 70% to mimic the surrounding soft tissues *in vivo* [58].

QCT has been proved to be a successful estimator for prediction of biomechanical properties of healing bones [7, 67]. Several experimental studies have verified its efficacy in such application. For instance, volumetric bone mineral density and cross-sectional area of fracture callus from an osteotomy model of sheep tibia with a defect size of 3 mm were measured using high resolution quantitative computed tomography with a slice thickness of 1 mm and an in-plane voxel size of 0.295 mm [67]. The value of the volumetric bone mineral density indicated a significant correlation with the bending stiffness of the fracture callus [67]. In addition, significant correlation between the torsional strength and QCT-derived bone strength was also observed in a rabbit tibial osteotomy model [7].

6.4.4 MICRO COMPUTED TOMOGRAPHY (MICRO-CT)

Micro computed tomography (micro-CT) is based on standard X-ray CT principles, but bone images are acquired with extremely high resolutions (up to 10μm). The high resolution of micro-CT enables the visualization of 3D micro-architectural distribution of trabecular bone, thus allowing for the detailed analysis of architectural parameters. Due to its high radiation doses, micro-CT is generally used in research environments and is not currently practical for use in clinical applications. The superior resolution of micro-CT makes itself a valuable tool in studying fracture healing in both small animals, such as rodents and rabbits [2, 4, 20, 22, 43, 68–74], and large animals, such as sheep [25, 75]. Micro-CT scans of fracture callus are normally done in a postmortem way due to the limitation of high radiation doses. Micro-CT can be used to measure structural parameters as total callus volume, mineralized callus volume, callus mineralized volume fraction (BV/TV), and compositional parameters, such as volumetric bone mineral density and mineral content in the fracture callus.

Biomechanical properties of repair bone can be predicted using certain architectural and compositional parameters of fracture callus derived from micro-CT images. The correlation between micro-CT prediction and the actual mechanical properties of callus has been verified in several studies [2, 68]. For example, investigators conducted both micro-CT imaging and torsional testing on fracture callus from mice femoral osteotomy models [2]. The results indicate that the torsional strength and rigidity of fracture callus correlate strongly with tissue mineral density, bone mineral content, and bone volume fraction (BV/TV) [2]. In another study of femoral transverse fractures in a rat model, micro-CT images and three-point bending test were performed on fracture calls [68]. Again, it was observed that bending strength and stiffness of fracture callus strongly correlated with the volume of mineralized callus and tissue mineral density [68]. However, both of the aforementioned studies indicate that the moment of inertia of the cross-section at the fracture callus is independent of its strength [2, 68].

6.5 NUMERICAL EVALUATION OF FRACTURE HEALING

Voxel-based finite element simulation has become a promising approach to predict mechanical properties of fracture callus in combination with micro-CT imaging techniques. With the micro-CT images and known bone mineral density distribution, the 3-D finite element model of fracture callus can be established to determine the mechanical behavior of healing bone. This method has been validated by comparing the predicted biomechanical properties of the fracture callus with those measured using three-point bending and torsional tests [4]. The predicted torsional rigidity of the callus correlated significantly with the experimental rigidity. However, neither of the callus area, bone mineral density or the moment of inertia showed such a correlation with the rigidity.

This approach has also been proved effective in clinical applications. For example, accurate predictions at the different healing stages of a tibia fracture after surgery of a 30-year old male patient were achieved using a 3D finite element model reconstructed from the micro-CT images, force plate data, and 3D interfragmentary micro-movements [76]. It was observed that load-sharing by the external fixator decreased significantly as the fracture tissue developed even moderate stiffness, while the load on the healing tibia increased steadily towards normal. Concurrently, the mechanical properties of the fracture callus at different stages of fracture healing can be calculated using the finite element model incorporated with the spatial distribution of bone mineral density determined by the micro-CT measurements.

6.6   SUMMARY

To allow comparison between different studies and their outcomes, it is essential that the methods of measuring biomechanical properties of tissue engineered/repair bone are standardized to ensure the acquisition of a reliable data pool that can serve as a data base for further developments of bone tissue engineering [77].

In pre-clinical animal models with segmental bone defects and non-critical size defects, torsional and bending tests have been extensively used for biomechanical evaluation of tissue engineered/repair bone samples. However, it is difficult to derive the intrinsic material properties of regenerated tissues from torsional and bending tests because these tests are structural tests and confounded with the geometry information of the fracture callus. Therefore, torsional and bending tests do not provide detailed information on how bone quality changes in the healing process. Recent advances in nanoindentation techniques represent a new direction in characterizing the quality of repair and tissue engineered bone as a function of location. As a result, spatial variation of biomechanical properties in fracture callus can be examined during the process of fracture healing.

In clinical application, plain x-ray images are used to assess the effect of healing of segmental defects as well as spontaneous healing. Although x-ray images provide qualitative information for fracture healing at different stages, these measurement cannot offer accurate characterization of the biomechanical properties of fracture callus. On the other hand, quantitative computed tomography can be used clinically to measure structure and composition of fracture callus and provide a successful estimation of biomechanical properties of healing bones. Consequently, quantitative computed tomography based analyses of fracture calluses could be a reliable tool for clinical assessment of fracture healing.

REFERENCES

[1] Liebschner, M.A., Biomechanical considerations of animal models used in tissue engineering of bone. Biomaterials, 2004. **25**(9): p. 1697–714. DOI: 10.1016/S0142-9612(03)00515-5 6.2

[2] Morgan, E.F., et al., Micro-computed tomography assessment of fracture healing: relationships among callus structure, composition, and mechanical function. Bone, 2009. **44**(2): p. 335–44. DOI: 10.1016/j.bone.2008.10.039 6.2.1, 6.4.4

[3] Ishihara, A., et al., Osteogenic gene regulation and relative acceleration of healing by adenoviral-mediated transfer of human BMP-2 or -6 in equine osteotomy and ostectomy models. J Orthop Res, 2008. **26**(6): p. 764–71. DOI: 10.1002/jor.20585 6.2.1, 6.2.1

[4] Shefelbine, S.J., et al., Prediction of fracture callus mechanical properties using micro-CT images and voxel-based finite element analysis. Bone, 2005. **36**(3): p. 480–8. DOI: 10.1016/j.bone.2004.11.007 6.2.1, 6.4.3, 6.4.4, 6.5

[5] Li, R.H., et al., rhBMP-2 injected in a calcium phosphate paste (alpha-BSM) accelerates healing in the rabbit ulnar osteotomy model. J Orthop Res, 2003. **21**(6): p. 997–1004. DOI: 10.1016/S0736-0266(03)00082-2 6.2.1, 6.2.1, 6.4.3

[6] Manigrasso, M.B. and J.P. O'Connor, Comparison of fracture healing among different inbred mouse strains. Calcif Tissue Int, 2008. **82**(6): p. 465–74. DOI: 10.1007/s00223-008-9144-3 6.2.1

[7] Chan, C.W., et al., Low intensity pulsed ultrasound accelerated bone remodeling during consolidation stage of distraction osteogenesis. J Orthop Res, 2006. **24**(2): p. 263–70. DOI: 10.1002/jor.20015 6.2.1, 6.2.1, 6.4.3

[8] Klein, P., et al., Comparison of unreamed nailing and external fixation of tibial diastases–mechanical conditions during healing and biological outcome. J Orthop Res, 2004. **22**(5): p. 1072–8. DOI: 10.1016/j.orthres.2004.02.006 6.2.1, 6.2.1

[9] Amanat, N., et al., A single systemic dose of pamidronate improves bone mineral content and accelerates restoration of strength in a rat model of fracture repair. J Orthop Res, 2005. **23**(5): p. 1029–34. DOI: 10.1016/j.orthres.2005.03.004 6.2.1, 6.2.1, 6.4.3

[10] Wang, H., et al., Thrombin peptide (TP508) promotes fracture repair by up-regulating inflammatory mediators, early growth factors, and increasing angiogenesis. J Orthop Res, 2005. **23**(3): p. 671–9. DOI: 10.1016/j.orthres.2004.10.002 6.2.1, 6.2.1

[11] Sammarco, G.J., et al., The biomechanics of torsional fractures: the effect of loading on ultimate properties. J Biomech, 1971. **4**(2): p. 113–7. DOI: 10.1016/0021-9290(71)90021-2 6.2.1

[12] Burstein, A.H. and V.H. Frankel, A standard test for laboratory animal bone. J Biomech, 1971. **4**(2): p. 155–8. DOI: 10.1016/0021-9290(71)90026-1 6.2.1

[13] White, A.A., 3rd, M.M. Panjabi, and W.O. Southwick, The four biomechanical stages of fracture repair. J Bone Joint Surg Am, 1977. **59**(2): p. 188–92. 6.2.1, 6.2.1

[14] Lienau, J., et al., CYR61 (CCN1) protein expression during fracture healing in an ovine tibial model and its relation to the mechanical fixation stability. J Orthop Res, 2006. **24**(2): p. 254–62. DOI: 10.1002/jor.20035 6.2.1

[15] Gerstenfeld, L.C., et al., Differential inhibition of fracture healing by non-selective and cyclooxygenase-2 selective non-steroidal anti-inflammatory drugs. J Orthop Res, 2003. **21**(4): p. 670–5. DOI: 10.1016/S0736-0266(03)00003-2 6.2.1

[16] Schell, H., et al., The course of bone healing is influenced by the initial shear fixation stability. J Orthop Res, 2005. **23**(5): p. 1022–8. DOI: 10.1016/j.orthres.2005.03.005 6.2.1

[17] Schemitsch, E.H., et al., Comparison of the effect of reamed and unreamed locked intramedullary nailing on blood flow in the callus and strength of union following fracture of the sheep tibia. J Orthop Res, 1995. **13**(3): p. 382–9. DOI: 10.1002/jor.1100130312 6.2.1

[18] Park, S.H. and M. Silva, Neuromuscular electrical stimulation enhances fracture healing: results of an animal model. J Orthop Res, 2004. **22**(2): p. 382–7. DOI: 10.1016/j.orthres.2003.08.007 6.2.1, 6.4.3

[19] Perka, C., et al., Segmental bone repair by tissue-engineered periosteal cell transplants with bioresorbable fleece and fibrin scaffolds in rabbits. Biomaterials, 2000. **21**(11): p. 1145–53. DOI: 10.1016/S0142-9612(99)00280-X 6.2.1, 6.4.1

[20] Komaki, H., et al., Repair of segmental bone defects in rabbit tibiae using a complex of beta-tricalcium phosphate, type I collagen, and fibroblast growth factor-2. Biomaterials, 2006. **27**(29): p. 5118–26. DOI: 10.1016/j.biomaterials.2006.05.031 6.2.1, 6.2.2, 6.4.1, 6.4.4

[21] Gao, T.J., et al., Enhanced healing of segmental tibial defects in sheep by a composite bone substitute composed of tricalcium phosphate cylinder, bone morphogenetic protein, and type IV collagen. J Biomed Mater Res, 1996. **32**(4): p. 505–12. DOI: 10.1002/(SICI)1097-4636(199612)32:4%3C505::AID-JBM2%3E3.0.CO;2-V 6.2.1, 6.4.1, 6.4.3, 6.4.3

[22] Oest, M.E., et al., Quantitative assessment of scaffold and growth factor-mediated repair of critically sized bone defects. J Orthop Res, 2007. **25**(7): p. 941–50. DOI: 10.1002/jor.20372 6.2.1, 6.4.1, 6.4.4

[23] Meinel, L., et al., Localized insulin-like growth factor I delivery to enhance new bone formation. Bone, 2003. **33**(4): p. 660–72. DOI: 10.1016/S8756-3282(03)00207-2 6.2.1

[24] Abendschein, W.F. and G.W. Hyatt, Ultrasonics and physical properties of healing bone. J Trauma, 1972. **12**(4): p. 297–301. DOI: 10.1097/00005373-197204000-00005 6.2.2

[25] Sarkar, M.R., et al., Bone formation in a long bone defect model using a platelet-rich plasma-loaded collagen scaffold. Biomaterials, 2006. **27**(9): p. 1817–23. DOI: 10.1016/j.biomaterials.2005.10.039 6.2.2, 6.4.1, 6.4.3, 6.4.4

[26] Augat, P., et al., Local tissue properties in bone healing: influence of size and stability of the osteotomy gap. J Orthop Res, 1998. **16**(4): p. 475–81. DOI: 10.1002/jor.1100160413 6.2.3, 6.2.5

[27] Jamsa, T., et al., Comparison of radiographic and pQCT analyses of healing rat tibial fractures. Calcif Tissue Int, 2000. **66**(4): p. 288–91. DOI: 10.1007/s002230010058 6.2.3, 6.2.4

[28] Luger, E.J., et al., Effect of low-power laser irradiation on the mechanical properties of bone fracture healing in rats. Lasers Surg Med, 1998. **22**(2): p. 97–102. DOI: 10.1002/(SICI)1096-9101(1998)22:2%3C97::AID-LSM5%3E3.0.CO;2-R 6.2.3

[29] Miclau, T., et al., Effects of delayed stabilization on fracture healing. J Orthop Res, 2007. **25**(12): p. 1552–8. DOI: 10.1002/jor.20435 6.2.3

[30] Paavolainen, P., et al., Calcitonin and fracture healing. An experimental study on rats. J Orthop Res, 1989. **7**(1): p. 100–6. DOI: 10.1002/jor.1100070114 6.2.3

[31] Prat, J., et al., Load transmission through the callus site with external fixation systems: theoretical and experimental analysis. J Biomech, 1994. **27**(4): p. 469–78. DOI: 10.1016/0021-9290(94)90022-1 6.2.3

[32] Rauch, F., et al., Effects of locally applied transforming growth factor-beta1 on distraction osteogenesis in a rabbit limb-lengthening model. Bone, 2000. **26**(6): p. 619–24. DOI: 10.1016/S8756-3282(00)00283-0 6.2.3

[33] Watanabe, Y., et al., Prediction of mechanical properties of healing fractures using acoustic emission. J Orthop Res, 2001. **19**(4): p. 548–53. DOI: 10.1016/S0736-0266(00)00042-5 6.2.3

[34] Black, J., et al., Stiffness and strength of fracture callus. Relative rates of mechanical maturation as evaluated by a uniaxial tensile test. Clin Orthop Relat Res, 1984(182): p. 278–88. 6.2.3

[35] Ekeland, A., L.B. Engesaeter, and N. Langeland, Mechanical properties of fractured and intact rat femora evaluated by bending, torsional and tensile tests. Acta Orthop Scand, 1981. **52**(6): p. 605–13. DOI: 10.3109/17453678108992155 6.2.3

[36] Dai, K.R., et al., Repairing of goat tibial bone defects with BMP-2 gene-modified tissue-engineered bone. Calcif Tissue Int, 2005. **77**(1): p. 55–61. DOI: 10.1007/s00223-004-0095-z 6.2.4

[37] Markel, M.D., et al., The determination of bone fracture properties by dual-energy X-ray absorptiometry and single-photon absorptiometry: a comparative study. Calcif Tissue Int, 1991. 48(6): p. 392–9. DOI: 10.1007/BF02556452 6.2.5, 6.4.2

[38] Markel, M.D., M.A. Wikenheiser, and E.Y. Chao, A study of fracture callus material properties: relationship to the torsional strength of bone. J Orthop Res, 1990. 8(6): p. 843–50. DOI: 10.1002/jor.1100080609 6.2.5

[39] Augat, P., L. Claes, and G. Suger, In vivo effect of shock-waves on the healing of fractured bone. Clin Biomech (Bristol, Avon), 1995. 10(7): p. 374–378. DOI: 10.1016/0268-0033(95)00009-A 6.2.5

[40] Manjubala, I., et al., Spatial and temporal variations of mechanical properties and mineral content of the external callus during bone healing. Bone, 2009. 45(2): p. 185–92. DOI: 10.1016/j.bone.2009.04.249 6.2.5

[41] Amanata, N., et al., The effect of zoledronic acid on the intrinsic material properties of healing bone: an indentation study. Med Eng Phys, 2008. 30(7): p. 843–7. DOI: 10.1016/j.medengphy.2007.09.008 6.2.5

[42] Leong, P.L. and E.F. Morgan, Measurement of fracture callus material properties via nanoindentation. Acta Biomaterialia, 2008. 4(5): p. 1569–1575. DOI: 10.1016/j.actbio.2008.02.030 6.2.5

[43] Leong, P.L. and E.F. Morgan, Correlations between indentation modulus and mineral density in bone-fracture calluses. Integrative and Comparative Biology, 2009. 49(1): p. 59–68. DOI: 10.1093/icb/icp024 6.2.5, 6.4.4

[44] Oliver, W.C. and G.M. Pharr, Measurement of hardness and elastic modulus by instrumented indentation: Advances in understanding and refinements to technology. J. Mater. Res, 2004. 2004(19): p. 1. 6.2.5

[45] Bourgois, R. and F. Burny, Measurement of the stiffness of fracture callus in vivo. A theoretical study. J Biomech, 1972. 5(1): p. 85–91. DOI: 10.1016/0021-9290(72)90021-8 6.3

[46] Lill, C.A., et al., Biomechanical evaluation of healing in a non-critical defect in a large animal model of osteoporosis. J Orthop Res, 2003. 21(5): p. 836–42. DOI: 10.1016/S0736-0266(02)00266-8 6.3, 6.4.3

[47] Thorey, F., et al., A new bending stiffness measurement device to monitor the influence of different intramedullar implants during healing period. Technol Health Care, 2008. 16(2): p. 129–40. 6.3, 6.4

[48] Hente, R., J. Cordey, and S.M. Perren, In vivo measurement of bending stiffness in fracture healing. Biomed Eng Online, 2003. 2: p. 8. DOI: 10.1186/1475-925X-2-8 6.3

[49] Maissen, O., et al., Mechanical and radiological assessment of the influence of rhTGFβ-3 on bone regeneration in a segmental defect in the ovine tibia: pilot study. J Orthop Res, 2006. **24**(8): p. 1670–8. DOI: 10.1002/jor.20231 6.3, 6.4.3

[50] Joslin, C.C., et al., Weight bearing after tibial fracture as a guide to healing. Clin Biomech (Bristol, Avon), 2008. **23**(3): p. 329–33. DOI: 10.1016/j.clinbiomech.2007.09.013 6.3

[51] Eastaugh-Waring, S.J., J.R. Hardy, and J.L. Cunningham, Fracture stiffness measurement using the orthometer: reproducibility and sources of error. Clin Biomech (Bristol, Avon), 2000. **15**(2): p. 140–2. DOI: 10.1016/S0268-0033(99)00029-7 6.3, 6.3

[52] Chotel, F., et al., Bone stiffness in children: part I. In vivo assessment of the stiffness of femur and tibia in children. J Pediatr Orthop, 2008. **28**(5): p. 534–7. 6.3

[53] Henry, A.N., M.A. Freeman, and S.A. Swanson, Studies on the mechanical properties of healing experimental fractures. Proc R Soc Med, 1968. **61**(9): p. 902–6. 6.3

[54] Davy, D.T. and J.F. Connolly, The biomechanical behavior of healing canine radii and ribs. J Biomech, 1982. **15**(4): p. 235–47. DOI: 10.1016/0021-9290(82)90170-1 6.3

[55] den Boer, F.C., et al., Healing of segmental bone defects with granular porous hydroxyapatite augmented with recombinant human osteogenic protein-1 or autologous bone marrow. J Orthop Res, 2003. **21**(3): p. 521–8. DOI: 10.1016/S0736-0266(02)00205-X 6.4.1

[56] Bloemers, F.W., et al., Autologous bone versus calcium-phosphate ceramics in treatment of experimental bone defects. J Biomed Mater Res B Appl Biomater, 2003. **66**(2): p. 526–31. DOI: 10.1002/jbm.b.10045 6.4.1

[57] Sheller, M.R., et al., Repair of rabbit segmental defects with the thrombin peptide, TP508. J Orthop Res, 2004. **22**(5): p. 1094–9. DOI: 10.1016/j.orthres.2004.03.009 6.4.1

[58] Blokhuis, T.J., et al., Resorbable calcium phosphate particles as a carrier material for bone marrow in an ovine segmental defect. J Biomed Mater Res, 2000. **51**(3): p. 369–75. DOI: 10.1002/1097-4636(20000905)51:3%3C369::AID-JBM10%3E3.0.CO;2-J 6.4.2, 6.4.3

[59] Gorman, S.C., et al., In vivo axial dynamization of canine tibial fractures using the Securos external skeletal fixation system. Vet Comp Orthop Traumatol, 2005. **18**(4): p. 199–207. 6.4.2

[60] den Boer, F.C., et al., New segmental long bone defect model in sheep: quantitative analysis of healing with dual energy x-ray absorptiometry. J Orthop Res, 1999. **17**(5): p. 654–60. DOI: 10.1002/jor.1100170506 6.4.2

[61] Markel, M.D. and J.J. Bogdanske, The effect of increasing gap width on localized densitometric changes within tibial ostectomies in a canine model. Calcif Tissue Int, 1994. **54**(2): p. 155–9. DOI: 10.1007/BF00296067 6.4.2

[62] Lee, S.C., et al., Healing of large segmental defects in rat femurs is aided by RhBMP-2 in PLGA matrix. J Biomed Mater Res, 1994. **28**(10): p. 1149–56. DOI: 10.1002/jbm.820281005 6.4.2

[63] Markel, M.D., et al., Atrophic nonunion can be predicted with dual energy x-ray absorptiometry in a canine ostectomy model. J Orthop Res, 1995. **13**(6): p. 869–75. DOI: 10.1002/jor.1100130610 6.4.2

[64] Cowin, S.C. Bone mechanics handbook, 2008, Informa. 6.4.2

[65] Schneiders, W., et al., In vivo effects of modification of hydroxyapatite/collagen composites with and without chondroitin sulphate on bone remodeling in the sheep tibia. J Orthop Res, 2009. **27**(1): p. 15–21. DOI: 10.1002/jor.20719 6.4.3

[66] Klein, P., et al., The initial phase of fracture healing is specifically sensitive to mechanical conditions. J Orthop Res, 2003. **21**(4): p. 662–9. DOI: 10.1016/S0736-0266(02)00259-0 6.4.3

[67] Augat, P., et al., Quantitative assessment of experimental fracture repair by peripheral computed tomography. Calcif Tissue Int, 1997. **60**(2): p. 194–9. DOI: 10.1007/s002239900213 6.4.3

[68] Nyman, J.S., et al., Quantitative measures of femoral fracture repair in rats derived by micro-computed tomography. Journal of Biomechanics, 2009. **42**(7): p. 891–897. DOI: 10.1016/j.jbiomech.2009.01.016 6.4.4

[69] Li, J.L., et al., P2X7 Nucleotide Receptor Plays an Important Role in Callus Remodeling During Fracture Repair. Calcified Tissue International, 2009. **84**(5): p. 405–412. DOI: 10.1007/s00223-009-9237-7 6.4.4

[70] Fu, L.J., et al., Effect of 1,25-dihydroxy vitamin D-3 on fracture healing and bone remodeling in ovariectomized rat femora. Bone, 2009. **44**(5): p. 893–898. DOI: 10.1016/j.bone.2009.01.378 6.4.4

[71] Warden, S.J., et al., Recombinant human parathyroid hormone (PTH 1–34) and low-intensity pulsed ultrasound have contrasting additive effects during fracture healing. Bone, 2009. **44**(3): p. 485–494. DOI: 10.1016/j.bone.2008.11.007 6.4.4

[72] Dickson, G.R., et al., Microcomputed tomography imaging in a rat model of delayed union/non-union fracture. Journal of Orthopaedic Research, 2008. **26**(5): p. 729–736. DOI: 10.1002/jor.20540 6.4.4

[73] Hao, Y.J., et al., Changes of microstructure and mineralized tissue in the middle and late phase of osteoporotic fracture healing in rats. Bone, 2007. **41**(4): p. 631–638. DOI: 10.1016/j.bone.2007.06.006 6.4.4

[74] Gabet, Y., et al., Osteogenic growth peptide modulates fracture callus structural and mechanical properties. Bone, 2004. **35**(1): p. 65–73. DOI: 10.1016/j.bone.2004.03.025 6.4.4

[75] Egermann, M., et al., Effect of BMP-2 gene transfer on bone healing in sheep. Gene Therapy, 2006. **13**(17): p. 1290–1299. DOI: 10.1038/sj.gt.3302785 6.4.4

[76] Vijayakumar, V., et al., Load transmission through a healing tibial fracture. Clin Biomech (Bristol, Avon), 2006. **21**(1): p. 49–53. DOI: 10.1016/j.clinbiomech.2005.08.011 6.5

[77] Reichert, J.C., et al., The challenge of establishing preclinical models for segmental bone defect research. Biomaterials, 2009. **30**(12): p. 2149–6. DOI: 10.1016/j.biomaterials.2008.12.050 6.6

CHAPTER 7

Mechanical and Structural Properties of Tissues Engineered/Repair Bone

CHAPTER SUMMARY

Chapter 7 provides a brief discussion on fundamentals of the self-repair process in fracture healing and intramembrane bone formation in scaffolds using the critical sized defect models. In addition, a large collection of experimental data is provided in this chapter regarding the mechanical and structural properties of tissue engineered/repair bone reported in the literature.

7.1 INTRODUCTION

The ultimate goal for implanting scaffolds to regenerate bone tissue is to fully repair the defect with the native tissue and restore mechanical integrity. As presented in the Chapter 6, there are a number of modalities for assessing the mechanical competence of a fracture callus. In this chapter, we present the biomechanical properties of repaired and regenerated tissues. These testing methods are typically performed *ex vivo* under various modes of loading, such as tension, compression, and torsion. Before presenting these properties as reported in the literature, the stages of fracture repair are described because the process of natural bone regeneration is informative to the design strategies of scaffolds for bone regeneration. For example, in the repair of a fracture, the regeneration process depends on mechanical strains that the callus experiences. Then, the critical sized defect model is described, followed by various biomechanical properties of regenerated tissue involving scaffolds.

7.2 BIOLOGICAL STAGES OF FRACTURE HEALING

Under normal conditions, bone has a relatively efficient repair mechanism, the biology of which continues to undergo extensive investigation [1]. Essentially, scaffolds help the repair process when factors are impeding bone from regenerating itself. Such factors include infection, large or critical-sized defect, and disease. Scaffold development must address these factors (e.g., deliver antibiotics) while promoting the repair mechanism. Moreover, scaffolds can hasten the repair process if they participate in a cooperative fashion with the biological mechanism of bone regeneration. What follows then is a description of the basic concepts of fracture repair.

Bone is capable of direct healing via osteonal reconstruction, but this repair process is not germane to scaffolds because it requires near absolute stability such that the two broken ends of the bone are in rigid contact [2]. The osteonal reconstruction requires an interfragmentary strain to be less than 2%. That is, any change in the fracture gap length normalized to the resting gap length must be quite small when bone is loaded. Scaffolds, especially porous ones, do not meet this requirement. Thus, they are typically developed to support secondary healing via endochondral bone formation (Figure 7.1) or direct healing via intramembraneous bone formation (Figure 7.2). Secondary healing

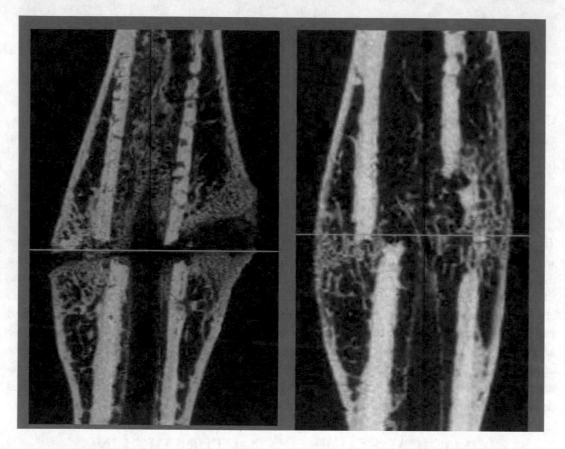

Figure 7.1: Secondary healing involves a process of endochondral ossification in which soft tissue fills the fracture forming a callus, calcifies, and then remodels. μCT images courtesy of Dr. Gregory Mundy and Dr. Gloria Gutierrez.

involves 4 biological phases that translate into 4 biomechanical stages. The biological phases are inflammatory, soft callus formation, callus mineralization, and remodeling.

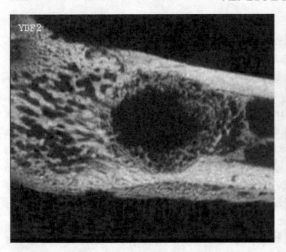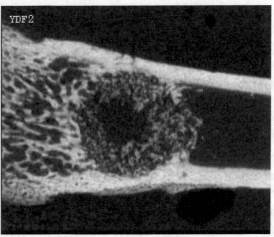

Figure 7.2: When a porous PUR scaffold was implanted into distal femur defect of rats, intramembraneous bone formation occurred directly as seen by μCT at 2 weeks (left) and 4 weeks (right). Images courtesy of Dr. Toshitaka Yoshii and Dr. Scott Guelcher.

7.2.1 INFLAMMATION

The first step in bone healing is the formation of a hematoma [3]. The fracture disrupts the periosteum, the connective tissue surrounding the bone, as well as the blood vessels. The broken vessels undergo thrombosis (clotting) as macrophages and leukoyctes invade to remove broken tissue fragments. The formation of new blood vessels or angiogenesis is essential for the deposition of the granulation tissue that initiates stabilization of the callus [4]. Not surprising then, the release of vascular endothelial growth factor (VEGF) from scaffolds has been studied as a means to promote bone regeneration [5–8]. The granulation tissue is relatively compliant and maintains its structure at high strains (>50%). The structural stiffness and strength of the callus are both low (Table 7.1) [9].

Table 7.1: Biomechanical properties of healing fracture change with each stage of repair (adapted from Bartel, Davy, Keaveny [9]).

Biological Stage	Stiffness	Strength	Location of failure
Inflammation	Low	Low	Fracture site
Soft tissue	High	Low	Fracture site
Calcified tissue	High	Moderate	Partial callus
Remodeling	High	High	Lowest cross-sectional area

7.2.2 SOFT TISSUE REPAIR

To increase the stability of the fracture site, granulation tissue is replaced with cartilage [10]. Pluripotential mesenchymal cells exist in both the periosteum and the bone marrow, and they differentiate into fibroblasts, chondrocytes, or osteoblasts. Thus, growth factors such as transforming growth factor beta (TGF-β) that increase this population of progenitor cells can cause a robust callus formation, especially if delivered early in the process of repair [11]. To hasten repair, however, factors such as fibroblast growth factors (FGFs) [12] or bone morphogenetic proteins (BMPs) [13] are needed to induce differentiation into the active mature cells that make connective tissue. Again, these factors can be incorporated into scaffolds [14–18], and ideally, would be released at the appropriate stage of repair. The cartilaginous tissue that bridges the fracture gap is stiffer than the interfragmentory tissue, and thus the fracture site is more stable at this stage with strains in the range of 10%. There is a direct relationship between strain level and callus size for small fracture gaps (less than 3 mm) [19]. At this stage, the structural stiffness is relatively high, but the callus strength is still low since the gap has not completely fused (Table 7.1).

7.2.3 CALCIFIED CARTILAGE FUSING THE FRACTURE GAP

Recapitulating the bone growth process known as endochondral ossification, the next stage involves cartilage mineralizing within the fracture site [20]. This of course further stiffens the callus and increases the stability of the fracture site. Bony union or fusion occurs when the mineralized tissue bridges the fracture gap as seen in μCT images (Figure 7.3). In humans, fusion occurs within 4 to 16 weeks with the process slowing with age [21–23]. At this point in the repair process, the callus strength is now close to the structural strength of non-fractured bone. The material properties of the callus tissue are not particularly strong or rigid though, but since the fracture gap is fused with a large callus having a diameter greater than an intact diaphysis, the fracture site can bear near physiological forces.

7.2.4 REMODELING OF THE CALLUS

In the final stage of fracture repair, the calcified cartilage is remodeled [24]. Monocytes fuse to form multi-nucleated osteoclasts that resorb the calcified cartilage, thereby decreasing the callus size. At first, osteoblasts rapidly form woven bone, but this tissue is eventually resorbed by the osteoclasts and replaced with lamellar bone in the form of osteons. During this phase, the callus strength can decrease as callus size decreases through remodeling [20]. However, as lamellar bone is deposited, the material properties increase returning the strength and stiffness to normal. From beginning to end of the natural repair of a fracture, stability is maintained initially by robust and large callus (structural response) that then returns to normal size (Figure 7.4) as woven bone is replaced by lamellar bone tissue.

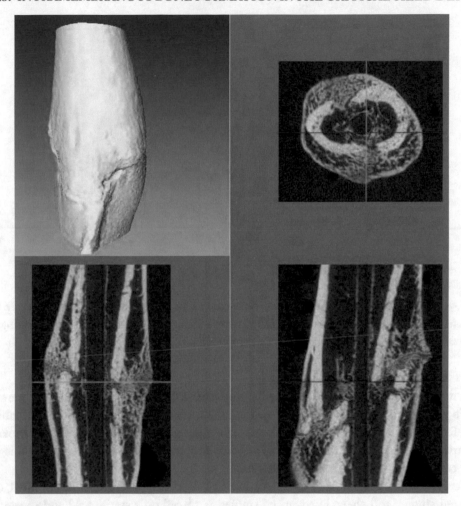

Figure 7.3: During fracture repair, the fracture gap is bridged as shown here by orthogonal, μCT images of a rat fracture callus after 4 weeks of healing. Images courtesy of Dr. Gregory Mundy and Dr. Gloria Gutierrez.

7.3 INTRAMEMBRANOUS BONE FORMATION IN THE CRITICAL SIZED DEFECT

To regenerate bone defects, scaffolds do not have to support all 4 stages of bone repair. While angiogenesis is important, the formation of soft tissue is not necessary because the scaffold itself is a surrogate acting to bridge the gap within the defect and provide stability. Bone tissue deposition can occur directly in a process called intramembraneous bone formation in which osteoblast secrete

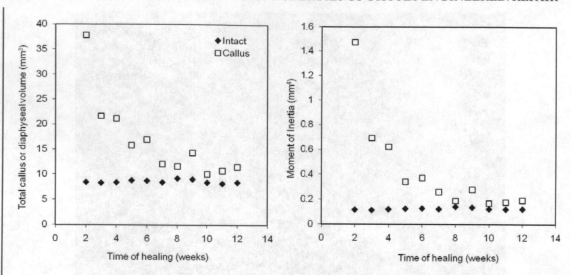

Figure 7.4: μCT-derived measurements show how the callus size for a mouse femur fracture typically decreases in a non-linear fashion to the size of the contra-lateral, intact femur during the healing process. Courtesy of Dr. Jonathan Schoenecker, Dr. Christopher Stutz, and Nicholas Mignemi.

osteoid on surfaces. The osteoid subsequently mineralizes. The critical sized defect is one situation when this process occurs with scaffolds providing the surface for bone deposition and stabilization of the defect. The scaffold is necessary because when the gap between two ends of a broken bone is of a particular size (greater than 3 mm), the bone does not heal. Therefore, a scaffold is inserted allowing cells to migrate and deposit tissue.

Here we describe a common model for testing the efficacy of scaffolds. The animal, typically rat in initial studies and sheep in advanced studies, is placed on a sterile field and covered with a surgical drape so that only the target limb is exposed. An incision is made to expose the bone of interest (e.g., diaphysis of a femur or tibia). The periosteum is removed from the bone, and a fixator template is fastened with two sterile cable ties. Next, a proximal hole is drilled with use of a drill guide and a sterile drill bit. The pin (e.g., a 1.1-mm threaded Kirschner wire in the case of rodents) is then secured to the bone. A distal pin is placed in the same fashion and is parallel to the proximal pin. Additional pins are placed as needed to provide stability. The template is then removed, and the skin is pulled over the pins. Small incisions are made to allow the pins to penetrate the skin. The external fixator is then secured as close to the skin as possible without risking skin ulcerations. An osteotomy is then performed to remove a segment of the diaphysis, this can be done using a sterile, round dental burr attached to a dental hand piece or a bone saw, a surgical instrument akin to a reciprocating saw. After completion of the osteotomy, the site is copiously irrigated with cefazolin solution. The prepared implants are inserted into the defect sites. The fascia is closed creating a tight muscle chamber around the defect. Bone regeneration can be monitored *in vivo* with weakly radiographs. Typical end-point measurements include bone volume as determined by μCT and torsional strength as determined by biomechanical testing.

7.4   BIOMECHANICAL PROPERTIES OF REPAIR/TISSUE ENGINEERED BONE

Information regarding biomechanical properties of repair/tissue engineered bone is important in two aspects. First, measuring biomechanical properties of fracture callus at various stages can help elucidate the underlying biomechanical principles of fracture healing. Second, characterizing biomechanical properties of healing bone can determine whether a certain tissue engineering strategy is successful in restoring the load-bearing capacity of fractured bone. Biomechanical properties of repair/tissue engineered bone have been measured under various loading modes. Consequently, torsional, bending, tensile, and compressive properties of repair/tissue engineered bone are presented.

Torsional properties of healing bone have been extensively reported in the literature. Bending properties of fracture callus are also available in a large number of studies. However, only a few studies have reported tensile and compressive properties of repair/tissue engineered bone. Most recently, compressive properties of healing bone have also been provided through micro/nanoindentation techniques.

7.4.1   TORSIONAL PROPERTIES

A large variation of torsional properties among repair/tissue engineered bone is observed due to different factors such as segmental defects vs. non-critical size defects, small animals vs. large animals, early stage vs. late stage of fracture healing.

Torsional properties of healing bone with segmental defects are available in a standard sheep tibia model to investigate the influence of various bone substitute materials and various bioactive agents on bone repair and regeneration [25–28]. In order to reduce errors associated with geometric and size differences among animal groups, torsional properties of fracture callus in the sheep tibia are expressed as a percentage of the value of the contralateral intact tibia (Table 7.2). Two types of tissue engineered bones, hydroxyapatite-based scaffolds and calcium phosphate-based constructs, have been employed in treating bone defects. Hydroxyapatite-based scaffolds are not fully resorbable whereas calcium phosphate-based scaffolds are resorbable during fracture healing [25,28]. Implantation of pure hydroxyapatite scaffolds into segmental bone defects yielded only limited new bone formation [27]. The recovery of torsional stiffness (18%) for bone defects implanted with pure hydroxyapatite is even less than that of empty defects (25%) (Table 7.2) [27]. Similarly, healing bone inserted with calcium phosphate particles has less recovery of torsional strength (6.9%) and stiffness (5.0%) than empty defects with a recovery of torsional strength (22%) and torsional stiffness (26%) (Table 7.2) [26]. Therefore, hydroxyapatite and calcium phosphate particles exhibit little osteoinductive activity [26,27].

Table 7.2: Comparison of strength and stiffness recovery among different treatment groups for segmental sheep defect model. (Data shown is the percentage recovery compared to that of the contralateral control bone.)

Author	Defect Size (mm)	Healing time (weeks)	Bone substitute material	Bioactive agents	Recovery of Torsional strength (%)	Recovery of Torsional stiffness (%)
Bloemers et al, 2003	30	12	Bone graft	N/A	40.5±7.5	48.0±12.2
			Hydroxyapatite particles	N/A	18.4±5.5	21.2±8.2
			Calcium phosphate particles	N/A	5.8±1.7	4.4±1.7
			Calcium phosphate paste	N/A	63.6±11.3	85.3±12.8
Den Boer et al, 2003	30	12	N/A	N/A	22	25
			Hydroxyapatite	N/A	19	18
			Hydroxyapatite	BMP-7	45	58
			Hydroxyapatite	Bone marrow	38	52
Blokhuis et al., 2000	30	12	Bone graft	N/A	40	48
			N/A	N/A	22±5	26±9
			Calcium Phosphate	N/A	6.9±1	5.0±1
			Calcium Phosphate	Bone marrow	26±7	20±6
Gao et al., 1996	16	12	Bone graft	N/A	41±8	48±12
			Calcium phosphate	N/A	63	80
			Calcium phosphate	13mg BMP	72	94
			Calcium phosphate	100mg BMP	137	125

On the other hand, a combination of bone substitutes and osteoinductive materials such as growth factors or autologous bone marrow may provide significant improvement for fracture healing. For example, osteogenic protein-1 (OP-1), one of the bone morphologic proteins, has shown potential in hastening fracture healing. In the sheep tibia model, torsional strength of healing bone injected with hydroxyapatite and recombinant OP-1 is two times higher than that of bone defects treated with hydroxyapatite alone or empty defects (Table 7.2) [27]. In another study, natural sheep bone morphologic proteins and a calcium phosphate cylinder are implanted into segmental bone defects of a sheep tibia model [28], resulting in a recovery of torsional strength (137%) and stiffness (125%) after 16 weeks of healing (Table 7.2). Furthermore, improvement of fracture healing is also observed with autologous bone marrow, which contains mesenchymal cells to differentiate into osteoblasts. Segmental bone defects filled with hydroxyapatite cylinders and soaked with autologous bone marrow have shown a higher strength (38%) and stiffness (52%) than that of empty defects (Table 7.2) [27].

Large animals such as sheep have been the choice of preclinical models for testing the effect of various tissue engineering strategies on bone regeneration. One of major reasons for this choice is that mature sheep have a body weight comparable to adult humans and long bone dimensions enabling the use of implants designed for the size of human bones. Nevertheless, small animals such as rats and rabbits remain popular and have also been used for evaluation of tissue engineering strategies on bone regeneration due to their low cost.

Torsional properties of healing bone with segmental defects are also available for small animal models such as rats [29] and rabbits [30]. In the rat model, 8-mm segmental defect model is created to evaluate the effect of polymeric scaffolds and growth factors on fracture healing [29]. After 16 weeks of treatment, little bone formation is observed in animals with empty defects. Consequently, torsional properties of healing bone with empty defects cannot be detected from mechanical testing. Compared to empty defects, polymeric scaffold-treated defects show a peak torque of 0.04 N·m, indicating an improvement of fracture healing. A combination of polymeric scaffolds and growth factors (BMP-7 and TGF-β3) further increases the torsional strength of healing bone to 0.06 N·m. However, the recovery of torsional strength from tissue engineer strategies is limited, compared to the strength of intact femur of rats (0.40 N·m). The low torsional strength of fracture callus can also be explained by its structure from micro-CT images (Figure 7.5). A segmental defect in rats does not form a union after 16 weeks of treatment. The limited restoration of load-bearing capacity of bone defects may be due to the presence of slow-degrading scaffolds or suboptimal doses of osteoinductive signals [29]. Nevertheless, the positive effect of scaffolds and bioactive agents has been observed in a rabbit ulna segmental defect model with a defect size of 15 mm [30]. The segmental defect at the left ulna is treated with polymeric microspheres only, whereas the right ulna is implanted with both polymeric microspheres and a synthetic peptide (TP508). Defects treated with $200\mu g$ TP508 have significantly higher torsional strength (0.29±0.16 N·m) and stiffness (0.033±0.013 N·m/deg.) than the ones without TP508 (torsional strength: 0.15±0.12 N·m, torsional stiffness: 0.026±0.012 N·m/deg.).

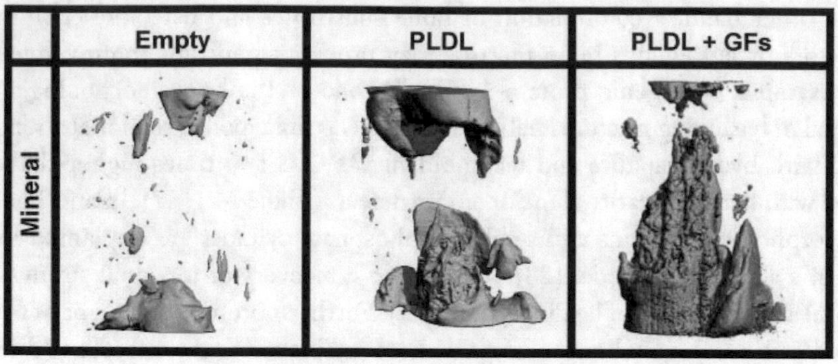

Figure 7.5: Micro-CT images of 8 mm segmental defects in rats for different treatment groups: empty defects; polymeric scaffolds; polymeric scaffolds and growth factors. (Reprinted from Oest et al., 2007, Journal of Orthopaedic Research 25:941–950.)

Animal models with non-critical size bone defects have been widely used to determine the effect of bioactive agents, and stimulation techniques on fracture healing.

Torsional properties of healing bone with non-critical size have been reported for small animals such as rats and rabbits [31–35] (Tables 7.3 and 7.4). Torsional strength of healing bone with empty defects in rat femurs ranges from 0.12 to 0.30 N·m (Table 7.3). Bioactive agents such as bisphosphonates, thrombin peptide (TP508), Vitamin D, BMP-2, BMP-7, PTH, and osteogenic growth peptide have positive effects on bone regeneration of non-critical size defects in rats and rabbits. Bone defects treated with bioactive agents show significantly higher torsional properties than the control group (Tables 7.3 and 7.4). In addition, ultrasound treatment [36] and neuromuscular electrical stimulation [37] in rabbit osteotomy models also demonstrate an improvement of torsional properties of healing bone.

7.4.2   BENDING PROPERTIES

Significant variations of bending properties are observed in repaired bone [38–42]. These variations may come from animal species, specimen geometry, testing conditions, defect sizes, and various tissue engineering strategies.

Bending properties of healing bone are available in mice, rats, rabbits, sheep, and goats. Small animals such as mice, rats and rabbits exhibit lower bending strength than large animals such as sheep and goats due to differences in specimen geometry. For instance, the bending strength of healing bone in rat femurs ranges from 65.15 to 133.4 N (Table 7.5).

Healing bone in rabbits exhibits bending strength of 281.2 N (Table 7.5). In a standard sheep tibia model, flexural rigidity of healing bone ranges from 2.45 N·m^2 for segmental defects to 57.5 N·m^2 for non-critical size defects (Table 7.5).

Table 7.3: List of torsional properties of healing bone among rat osteotomy models.

Author	Animal models	Healing Time (weeks)	Treatment groups	Torsional strength (N·m)	Torsional stiffness (N·m/deg)	Energy to failure (N·m·deg)
Amanat et al., 2005 Rat ulna		6	Control	0.35±0.11	0.017±0.01	4.48±2.0
			Local Bisphosphonate	0.41±0.21	0.027±0.01	4.50±4.4
			Systemic Bisphosphonate	0.56±0.25	0.039±0.02	5.15±3.0
Wang et al., 2005	Rat femur	4	Control	0.3	N/A	N/A
			Thrombin peptide (TP508)	0.41	N/A	N/A
Gabet et al., 2004	Rat femur	4	Control	0.13	0.56 N.m/rad	0.022 N.m.rad
			Osteogenic growth peptide	0.14	0.57 N.m/rad	0.028 N.m.rad
Delgado-Martinez et al., 1998	Rat femur	5	Control	0.12±0.032	0.010±0.005 N.m/rad	0.020 J N.m.rad
			Vitamin D	0.181±0.043	0.015±0.007	0.038 J

Table 7.4: List of torsional properties of healing bone among rabbit osteotomy models.

Author	Animal models	Healing Time (weeks)	Treatment groups	Torsional strength (N·m)	Torsional stiffness (N·m/deg)	Energy to failure (N·m·deg)
Morgan et al., 2008	Rabbit tibia	4	BMP-7 and PTH	N/A	0.65	N/A
		4	BMP-7	N/A	0.50	N/A
		4	PTH	N/A	0.52	N/A
		4	Control	N/A	0.48	N/A
Li et al, 2003	Rabbit ulna		BMP-2	0.65±0.13	0.031±0.01	8.36±1.6
			Control	0.25±0.13	0.011±0.007	3.03±1.7
White et al., 1977	Rabbit tibia	4	Control	0.35±0.31	0.081±0.106	0.016±0.007

Table 7.5: Bending properties of fracture callus.

Author	Animal models	Healing time (weeks)	Treatment groups	Bending strength (N)	Bending stiffness (N/mm)	Energy to failure (N·mm)
Nyman et al., 2009	Rat Femur Closed transverse osteotomy	4	Empty defects	90.6±22.0	140±85	37.2±13.0
			lovasatin	133.4±15.9	196±87	56.2±17.9
Fu et al., 2009	Rat femur Transverse osteotomy	16	Empty defects	65.12±3.46	N/A	88.20±3.08
			Vitamin D3	72.82±3.20	N/A	99.23±4.17
Dickson et al., 2008	Rat femur Transverse osteotomy	14	Empty defects	81.72±57.94	N/A	N/A
			Bone marrow cells	100.34±43.74	N/A	N/A
Sarkar et al., 2006	Sheep tibia 25 mm Segmental defects	12	Empty defects	N/A	2.45 Nm2 (flexural rigidity)	N/A
			Platelet-rich plasma	N/A	3.41 Nm2	N/A
Komaki et al., 2006	Rabbit tibia 5mm segmental defects	12	Empty	N/A	N/A	N/A
			Calcium phosphate	281.2±57	N/A	N/A
			Calcium phosphate, FGF-2	583.9±116	N/A	N/A

Bending properties of healing bone from the same animal model may be different due to changes in testing conditions. For example, in a rat femur transverse osteotomy model, a three-point bending test [40] resulted in a lower bending strength of healing bone than a four-point bending test [39]. This may be explained by the configuration difference between three-point and four-point bending tests.

Bending properties of fracture callus with segmental defects are significantly improved through the use of scaffolds and growth factors. For example, in a rabbit tibia model with segmental defects, bending properties were not available for healing bone without scaffold since there was no union in the fracture callus [41]. After a treatment of calcium phosphate scaffolds, the bending strength of healing was 281.2 N. With a combination of scaffolds and growth factors, the bending strength of healing of fracture callus was doubled (Table 7.5).

Bending properties of fracture callus with non-critical size defects also benefit from the use of tissue engineering strategies [38–40]. As an example, empty defects with non-critical size defects in rat femur can achieve a bending strength of 90.6 N after 4 weeks of fracture healing [38]. With the treatment of lovasatin, the bending strength of healing increased to 133.4 N (Table 7.5).

7.4.3   TENSILE PROPERTIES

Most of the tensile properties of fracture callus have been reported for small animal models such as mice [43], rats [44–46], and rabbits [47] (Table 7.6).

Tensile strength of fracture callus for rat femoral fracture models ranges from 36N at early stage of facture healing (4 weeks) to 510N at the late stage (10 weeks) [44]. Tensile properties of fracture callus have been found to correlate with callus size, bone mineral content, and bone mineral density [45].

Tensile properties of fracture callus were higher in the treatment with laser irradiation than the group without treatment [46]. On the other hand, in another study, tensile properties of fracture callus treated with TGF-β1 were not significantly different from the control group, suggesting that TGF-β1 may not have an effective treatment for fracture healing [47].

7.4.4   COMPRESSIVE PROPERTIES OF FRACTURE CALLUS

Only a few studies have reported compressive properties of facture callus [45, 48–50]. Compressive properties of healing bone are available for segmental bone defects and non-critical size defects. Associations between compressive properties of fracture callus and bone mineral density and callus size have also been observed.

In critical-size defects of goat tibial models, a combination of biphasic calcined bone, autologous bone marrow and bone morphologic protein-2 have been investigated in the treatment of fracture healing [50]. Compressive strength (26.769±7.791 MPa) of fracture callus after 26 weeks of implantation was three times higher than the group without BMP-2 treatment.

A rabbit tibia model with a wedge shape defect was used to test the effect of bone morphogenetic protein 7 (BMP-7) and parathyroid hormone (PTH) on fracture healing [48]. Combined

Table 7.6: Tensile properties of fracture callus.

Author	Animal models	Healing time (weeks)	Treatment groups	Tensile strength (N)	Tensile stiffness (N/mm)	Energy to failure (N·mm)
Miclau et al., 2007	Mice tibia Closed transverse fractures	2	Empty defects	5.09±1.33	4.51±1.81	N/A
Watanabe et al., 2001	Rat femur 2mm	4	Empty defects	36±21	0.47±0.18	N/A
		6		130±60	1.3±6	N/A
		8		222±32	1.8±6	N/A
		10		510±95	3.0±2	N/A
Jamas et al., 2000	Rat tibia Transverse fracture	4	Empty defects	N/A	N/A	N/A
Luger et al., 1998	Rat tibia Transverse fracture	4	Empty defects	46.5±20.2	13.97±10.27	N/A
			Laser irradiation	74.4±43.1	26.28±11.88	N/A
Rauch et al.,2000	Rabbit tibia osteotomy	4	Empty defects	850±160	14100±6200	N/A
			TGF-b1	670±180	6600±2900	N/A

treatment with BMP-7 and PTH resulted in higher compressive strength compared to the control group (Figure 7.6). Compressive strength of fracture callus was higher in the PTH group than in the BMP-7 group [48].

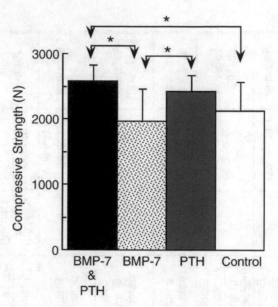

Figure 7.6: Comparison of compressive strength of fracture callus from a rabbit tibia fracture model with different treatment groups. (Reprinted from Morgan et al., 2008, Bone 43:1031–1038.)

Compressive strength of fracture callus in a rat tibial fracture model was associated with the callus size, bone mineral density, and bone mineral density measured by peripheral quantitative computed tomography [45].

In addition, compressive strain distribution of fracture callus was mapped by electronic speckle pattern interferometry for a sheep tibial fracture model [49]. The compressive stiffness of fracture callus was 10.3 kN/mm after 8 weeks of fracture healing [49]. The compressive strain distribution of fracture callus in Figure 7.7 indicated that the highest strain was detected adjacent to cortical boundaries in the osteotomy gap.

Compressive properties of fracture callus have also been assessed by indentation tests. Indentation stiffness of fracture callus changes during the course of fracture healing. In a canine tibial fracture model, indentation stiffness changes from 12.7 N/mm at 2 weeks to 153.7 N/mm at 12 weeks [51]. In a sheep tibial osteotomy model, indentation modulus of fracture callus changes from 6 GPa at early stage to 14 GPa at a later stage (Figure 7.8) [52].

Spatial variation of indentation stiffness has been observed among different regions of fracture callus. For example, indentation modulus ranged from 0.51 MPa to 1680 MPa throughout the fracture callus of rat femoral osteotomy models [53]. Indentation stiffness of fracture callus have been

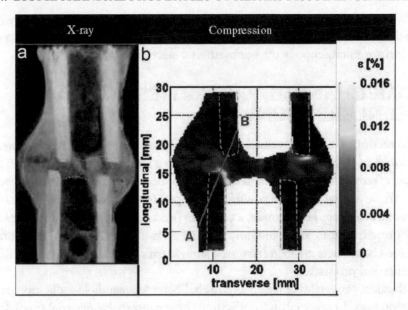

Figure 7.7: (a) Contact radiograph of fracture callus; (b) Strain distribution fracture callus under compression. (Reprinted from Bottland et al., 2008, Journal of Biomechanics 41:701–705.)

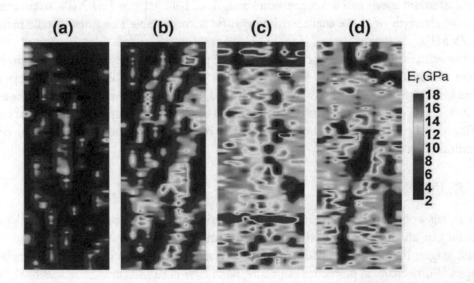

Figure 7.8: Maps of indentation modulus for fracture callus at different stages of fracture healing: (a) 2 weeks, (b) 3 weeks, (c) 6 weeks, and (d) 9 weeks. (Reprinted from Manjubala et al., 2009, Bone 45:185–192.)

shown to associate with bone mineral content measured by dual-energy X-ray absorptiometry [54], tissue mineral density measured by micro-computed tomography [53], calcium content characterized by scanning electron microscopy in the backscattered electron mode [52].

7.5 BIOMECHANICAL PROPERTIES OF TISSUE ENGINEERED BONE

In the previous section, we have described the biomechanical properties of repair bone. Repair bone is mostly produced from in vivo animal models in which a bone defect is first created and scaffolds, cells and growth factors are then implanted into the bone defect for accelerating fracture healing. In other cases, tissue engineered bone could be developed from in vivo animal models without bone defects and fracture healing. For example, a common practice is available to implant scaffolds filled with cells and growth factors subcutaneously in nude mice [55] and rats [56], and infant pigs [57]. Tissue engineered constructs are then harvested from sacrificed animals and mechanically tested to obtain biomechanical properties.

Biomechanical properties of tissue engineered bone for mandible replacements are available from compression tests. In one study, scaffolds from either a resorbable material as polylactite mesh or a non-resorbable material as titanium mesh with recombinant human bone morphogenetic protein-7 were implanted into muscles of infant minipigs [57]. After 6 weeks, the mandible replacements were harvested and tested under compression (Figure 7.9). Tissue engineered constructs with a polyactite mesh or a titanium mesh had a compression strength of 1.62 MPa or 1.51 MPa, respectively. The compressive strength of tissue engineered constructs is comparable to natural porcile mandibular bone: 1.75 MPa.

Biomechanical properties of tissue engineered constructs are dependent on the time of implantation within the host animal. When tricalcium phosphate scaffolds and mesenchymal stem cells were implanted subcutaneously in nude mice [55], tissue engineered constructs were acquired after 1, 2, 4, and 6 weeks of implantation in mice and tested under compression to obtain compressive stiffness. All constructs demonstrated gradually increasing stiffness during the time course of implantation (Figure 7.10).

7.6 SUMMARY

Biological stages of fracture healing include inflammation, soft tissue repair, calcified cartilage fusing the fracture gap, and remodeling of the callus. From mechanical testing of fracture callus at different biological stages, biomechanical properties of the fracture callus also have been categorized into four stages. Biomechanical properties of healing bone have been measured from torsional, bending, tensile and compressive tests. When tissue engineering strategies such as incorporating cells and/or growth factors into a scaffold are applied in fracture healing, biomechanical properties of repair bone can be significantly enhanced. In addition to repair bone, the development of tissue engineered bone is also available with the use of bioreactors, bone chambers, and in vivo animal models. Although

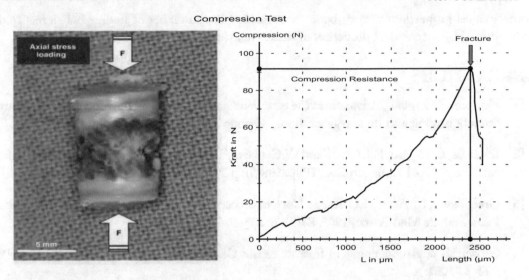

Figure 7.9: Tissue engineered bone for mandible replacements (left) is loaded axially into a compression test device. Compression stress (right) increases gradually during the course of loading. (Reprinted from Warnke et al., 2006, Biomaterials 7: 1081-1087.)

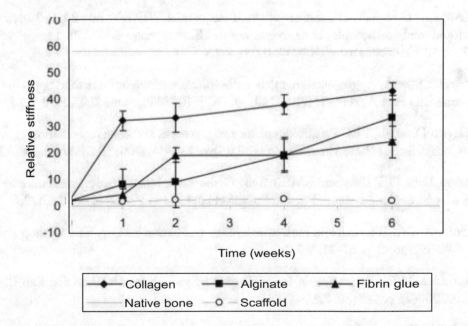

Figure 7.10: Changes of relative compressive stiffness for tissue engineered constructs with implantation time. (Reprinted from Weinand et al., 2007, Tissue Engineering 13: 757-765.)

biomechanical properties of repair bone are available from a number of studies, only a few studies have reported biomechanical properties of tissue engineered bone from in vivo animal studies.

REFERENCES

[1] Holstein, J.H., et al., Advances in the establishment of defined mouse models for the study of fracture healing and bone regeneration. J Orthop Trauma, 2009. **23**(5 Suppl): p. S31–8. 7.2

[2] DiGioia, A.M., 3rd, E.J. Cheal, and W.C. Hayes, Three-dimensional strain fields in a uniform osteotomy gap. J Biomech Eng, 1986. **108**(3): p. 273–80. DOI: 10.1115/1.3138614 7.2

[3] Greenbaum, M.A. and I.O. Kanat, Current concepts in bone healing. Review of the literature. J Am Podiatr Med Assoc, 1993. **83**(3): p. 123–9. 7.2.1

[4] Glowacki, J., Angiogenesis in fracture repair. Clin Orthop Relat Res, 1998(355 Suppl): p. S82–9. 7.2.1

[5] Chen, L., et al., Loading of VEGF to the heparin cross-linked demineralized bone matrix improves vascularization of the scaffold. J Mater Sci Mater Med, 2009. DOI: 10.1007/s10856-009-3827-9 7.2.1

[6] Kempen, D.H., et al., Effect of local sequential VEGF and BMP-2 delivery on ectopic and orthotopic bone regeneration. Biomaterials, 2009. **30**(14): p. 2816–25. DOI: 10.1016/j.biomaterials.2009.01.031 7.2.1

[7] Barralet, J., et al., Angiogenesis in calcium phosphate scaffolds by inorganic copper ion release. Tissue Eng Part A, 2009. **15**(7): p. 1601–9. DOI: 10.1089/ten.tea.2007.0370 7.2.1

[8] Kaigler, D., et al., VEGF scaffolds enhance angiogenesis and bone regeneration in irradiated osseous defects. J Bone Miner Res, 2006. **21**(5): p. 735–44. DOI: 10.1359/jbmr.060120 7.2.1

[9] Bartel, D.L., D.T. Davy, and T.M. Keaveny, Orthopaedic biomechanics: mechanics and design in musculoskeletal systems. Pearson Prentice Hall bioengineering. 2006, 7.2.1, 7.1

[10] Einhorn, T.A., The cell and molecular biology of fracture healing. Clin Orthop Relat Res, 1998(355 Suppl): p. S7–21. 7.2.2

[11] Bolander, M.E., Regulation of fracture repair by growth factors. Proc Soc Exp Biol Med, 1992. **200**(2): p. 165–70. 7.2.2

[12] Nakajima, F., et al., Effects of a single percutaneous injection of basic fibroblast growth factor on the healing of a closed femoral shaft fracture in the rat. Calcif Tissue Int, 2007. **81**(2): p. 132–8. DOI: 10.1007/s00223-007-9048-7 7.2.2

[13] Einhorn, T.A., et al., A single percutaneous injection of recombinant human bone morphogenetic protein-2 accelerates fracture repair. J Bone Joint Surg Am, 2003. **85-A**(8): p. 1425–35. 7.2.2

[14] Liu, Y., et al., Segmental bone regeneration using an rhBMP-2-loaded gelatin/nanohydroxyapatite/fibrin scaffold in a rabbit model. Biomaterials, 2009. **30**(31): p. 6276–85. DOI: 10.1016/j.biomaterials.2009.08.003 7.2.2

[15] Yilgor, P., N. Hasirci, and V. Hasirci, Sequential BMP-2/BMP-7 delivery from polyester nanocapsules. J Biomed Mater Res A, 2009. DOI: 10.1002/jbm.a.32520 7.2.2

[16] Yilgor, P., et al., Incorporation of a sequential BMP-2/BMP-7 delivery system into chitosan-based scaffolds for bone tissue engineering. Biomaterials, 2009. **30**(21): p. 3551–9. DOI: 10.1016/j.biomaterials.2009.03.024 7.2.2

[17] Keiichi, K., et al., Induction of new bone by basic FGF-loaded porous carbonate apatite implants in femur defects in rats. Clin Oral Implants Res, 2009. **20**(6): p. 560–5. 7.2.2

[18] Gomez, G., et al., Effect of FGF and polylactide scaffolds on calvarial bone healing with growth factor on biodegradable polymer scaffolds. J Craniofac Surg, 2006. **17**(5): p. 935–42. DOI: 10.1097/01.scs.0000231624.87640.55 7.2.2

[19] Claes, L., et al., Influence of size and stability of the osteotomy gap on the success of fracture healing. J Orthop Res, 1997. **15**(4): p. 577–84. DOI: 10.1002/jor.1100150414 7.2.2

[20] Martin, R.B., D.B. Burr, and N.A. Sharkey, Skeletal Tissue Mechanics. 1998, Springer-Verlag. DOI: 10.1016/S0021-9290(00)00029-4 7.2.3, 7.2.4

[21] Bak, B. and T.T. Andreassen, The effect of aging on fracture healing in the rat. Calcif Tissue Int, 1989. **45**(5): p. 292–7. DOI: 10.1007/BF02556022 7.2.3

[22] Meyer, R.A., Jr., et al., Age and ovariectomy impair both the normalization of mechanical properties and the accretion of mineral by the fracture callus in rats. J Orthop Res, 2001. **19**(3): p. 428–35. DOI: 10.1016/S0736-0266(00)90034-2 7.2.3

[23] Lu, C., et al., Cellular basis for age-related changes in fracture repair. J Orthop Res, 2005. **23**(6): p. 1300–7. DOI: 10.1016/j.orthres.2005.04.003.1100230610 7.2.3

[24] Crenshaw, A.H., K. Daugherty, and W.C. Campbell, Campbell's operative orthopaedics. Mosby Year Book. 7.2.4

[25] Bloemers, F.W., et al., Autologous bone versus calcium-phosphate ceramics in treatment of experimental bone defects. J Biomed Mater Res B Appl Biomater, 2003. **66**(2): p. 526–31. DOI: 10.1002/jbm.b.10045 7.4.1, 7.4.1

[26] Blokhuis, T.J., et al., Resorbable calcium phosphate particles as a carrier material for bone marrow in an ovine segmental defect. J Biomed Mater Res, 2000. **51**(3): p. 369–75. DOI: 10.1002/1097-4636(20000905)51:3%3C369::AID-JBM10%3E3.0.CO;2-J 7.4.1, 7.4.1

[27] den Boer, F.C., et al., Healing of segmental bone defects with granular porous hydroxyapatite augmented with recombinant human osteogenic protein-1 or autologous bone marrow. J Orthop Res, 2003. **21**(3): p. 521–8. DOI: 10.1016/S0736-0266(02)00205-X 7.4.1, 7.4.1

[28] Gao, T.J., et al., Enhanced healing of segmental tibial defects in sheep by a composite bone substitute composed of tricalcium phosphate cylinder, bone morphogenetic protein, and type IV collagen. J Biomed Mater Res, 1996. **32**(4): p. 505–12. DOI: 10.1002/(SICI)1097-4636(199612)32:4%3C505::AID-JBM2%3E3.0.CO;2-V 7.4.1, 7.4.1

[29] Oest, M.E., et al., Quantitative assessment of scaffold and growth factor-mediated repair of critically sized bone defects. J Orthop Res, 2007. **25**(7): p. 941–50. DOI: 10.1002/jor.20372 7.4.1, 7.4.1

[30] Sheller, M.R., et al., Repair of rabbit segmental defects with the thrombin peptide, TP508. J Orthop Res, 2004. **22**(5): p. 1094–9. DOI: 10.1016/j.orthres.2004.03.009 7.4.1, 7.4.1

[31] Morgan, E.F., et al., Micro-computed tomography assessment of fracture healing: relationships among callus structure, composition, and mechanical function. Bone, 2009. **44**(2): p. 335–44. DOI: 10.1016/j.bone.2008.10.039 7.4.1

[32] Wang, H., et al., Thrombin peptide (TP508) promotes fracture repair by up-regulating inflammatory mediators, early growth factors, and increasing angiogenesis. J Orthop Res, 2005. **23**(3): p. 671–9. DOI: 10.1016/j.orthres.2004.10.002 7.4.1

[33] Gabet, Y., et al., Osteogenic growth peptide modulates fracture callus structural and mechanical properties. Bone, 2004. **35**(1): p. 65–73. DOI: 10.1016/j.bone.2004.03.025 7.4.1

[34] Delgado-Martinez, A.D., et al., Effect of 25-OH-vitamin D on fracture healing in elderly rats. J Orthop Res, 1998. **16**(6): p. 650–3. DOI: 10.1002/jor.1100160604 7.4.1

[35] Amanat, N., et al., A single systemic dose of pamidronate improves bone mineral content and accelerates restoration of strength in a rat model of fracture repair. J Orthop Res, 2005. **23**(5): p. 1029–34. DOI: 10.1016/j.orthres.2005.03.004 7.4.1

[36] Chan, C.W., et al., Low intensity pulsed ultrasound accelerated bone remodeling during consolidation stage of distraction osteogenesis. J Orthop Res, 2006. **24**(2): p. 263–70. DOI: 10.1002/jor.20015 7.4.1

[37] Park, S.H. and M. Silva, Neuromuscular electrical stimulation enhances fracture healing: results of an animal model. J Orthop Res, 2004. **22**(2): p. 382–7. DOI: 10.1016/j.orthres.2003.08.007 7.4.1

[38] Nyman, J.S., et al., Quantitative measures of femoral fracture repair in rats derived by micro-computed tomography. Journal of Biomechanics, 2009. **42**(7): p. 891–897. DOI: 10.1016/j.jbiomech.2009.01.016 7.4.2, 7.4.2

[39] Dickson, G.R., et al., Microcomputed tomography imaging in a rat model of delayed union/non-union fracture. Journal of Orthopaedic Research, 2008. **26**(5): p. 729–736. DOI: 10.1002/jor.20540 7.4.2, 7.4.2

[40] Fu, L.J., et al., Effect of 1,25-dihydroxy vitamin D-3 on fracture healing and bone remodeling in ovariectomized rat femora. Bone, 2009. **44**(5): p. 893–898. DOI: 10.1016/j.bone.2009.01.378 7.4.2, 7.4.2

[41] Komaki, H., et al., Repair of segmental bone defects in rabbit tibiae using a complex of beta-tricalcium phosphate, type I collagen, and fibroblast growth factor-2. Biomaterials, 2006. **27**(29): p. 5118–26. DOI: 10.1016/j.biomaterials.2006.05.031 7.4.2, 7.4.2

[42] Sarkar, M.R., et al., Bone formation in a long bone defect model using a platelet-rich plasma-loaded collagen scaffold. Biomaterials, 2006. **27**(9): p. 1817–23. DOI: 10.1016/j.biomaterials.2005.10.039 7.4.2

[43] Miclau, T., et al., Effects of delayed stabilization on fracture healing. J Orthop Res, 2007. **25**(12): p. 1552–8. DOI: 10.1002/jor.20435 7.4.3

[44] Watanabe, Y., et al., Prediction of mechanical properties of healing fractures using acoustic emission. J Orthop Res, 2001. **19**(4): p. 548–53. DOI: 10.1016/S0736-0266(00)00042-5 7.4.3, 7.4.3

[45] Jamsa, T., et al., Comparison of radiographic and pQCT analyses of healing rat tibial fractures. Calcif Tissue Int, 2000. **66**(4): p. 288–91. DOI: 10.1007/s002230010058 7.4.3, 7.4.3, 7.4.4, 7.4.4

[46] Luger, E.J., et al., Effect of low-power laser irradiation on the mechanical properties of bone fracture healing in rats. Lasers Surg Med, 1998. **22**(2): p. 97–102. DOI: 10.1002/(SICI)1096-9101(1998)22:2%3C97::AID-LSM5%3E3.0.CO;2-R 7.4.3, 7.4.3

[47] Rauch, F., et al., Effects of locally applied transforming growth factor-beta1 on distraction osteogenesis in a rabbit limb-lengthening model. Bone, 2000. **26**(6): p. 619–24. DOI: 10.1016/S8756-3282(00)00283-0 7.4.3, 7.4.3

[48] Morgan, E.F., et al., Combined effects of recombinant human BMP-7 (rhBMP-7) and parathyroid hormone (1–34) in metaphyseal bone healing. Bone, 2008. **43**(6): p. 1031–8. DOI: 10.1016/j.bone.2008.07.251 7.4.4

[49] Bottlang, M., et al., Acquisition of full-field strain distributions on ovine fracture callus cross-sections with electronic speckle pattern interferometry. J Biomech, 2008. **41**(3): p. 701–5. DOI: 10.1016/j.jbiomech.2007.10.024 7.4.4, 7.4.4

[50] Dai, K.R., et al., Repairing of goat tibial bone defects with BMP-2 gene-modified tissue-engineered bone. Calcif Tissue Int, 2005. **77**(1): p. 55–61. DOI: 10.1007/s00223-004-0095-z 7.4.4

[51] Markel, M.D., M.A. Wikenheiser, and E.Y. Chao, A study of fracture callus material properties: relationship to the torsional strength of bone. J Orthop Res, 1990. **8**(6): p. 843–50. DOI: 10.1002/jor.1100080609 7.4.4

[52] Manjubala, I., et al., Spatial and temporal variations of mechanical properties and mineral content of the external callus during bone healing. Bone, 2009. **45**(2): p. 185–92. DOI: 10.1016/j.bone.2009.04.249 7.4.4

[53] Leong, P.L. and E.F. Morgan, Correlations between indentation modulus and mineral density in bone-fracture calluses. Integrative and Comparative Biology, 2009. **49**(1): p. 59–68. DOI: 10.1093/icb/icp024 7.4.4

[54] Markel, M.D., et al., The determination of bone fracture properties by dual-energy X-ray absorptiometry and single-photon absorptiometry: a comparative study. Calcif Tissue Int, 1991. **48**(6): p. 392–9. DOI: 10.1007/BF02556452 7.4.4

[55] Weinand, C., et al., Comparison of hydrogels in the in vivo formation of tissue-engineered bone using mesenchymal stem cells and beta-tricalcium phosphate. Tissue Eng, 2007. **13**(4): p. 757–65. DOI: 10.1089/ten.2006.0083 7.5, 7.5

[56] Zhou, Y., et al., Combined marrow stromal cell-sheet techniques and high-strength biodegradable composite scaffolds for engineered functional bone grafts. Biomaterials, 2007. **28**(5): p. 814–24. DOI: 10.1016/j.biomaterials.2006.09.032 7.5

[57] Warnke, P.H., et al., The mechanical integrity of in vivo engineered heterotopic bone. Biomaterials, 2006. **27**(7): p. 1081–7. DOI: 10.1016/j.biomaterials.2005.07.042 7.5

CHAPTER 8

Current Issues of Biomechanics in Bone Tissue Engineering

CHAPTER SUMMARY

After several decades of endeavor, tissue engineers and researchers have realized that synergetic efforts from multiple disciplines in engineering, biology, physics, chemistry, and other related fields are imperative for successful tissue regeneration. In response to such a demand, the concept of functional tissue engineering has been proposed to make an implant system mimicking the natural environment for healthy tissue restoration and regeneration. Currently, the development of ideal scaffold systems that possess not only structural but also biological, biophysical functions is required. In addition, design and fabrication of novel bioreactors have become a new challenge to mimic the *in vivo* environment of tissue regeneration. Finally, the elucidation of the underlying mechanism of bone mechanobiology is required in order for bioengineers to develop criteria in design and fabrication of implant systems. This chapter is intended to briefly discuss these aspects, respectively.

8.1 INTRODUCTION

The current challenge in bone tissue engineering is the development of ideal scaffold-cell systems and the establishment of an adequate environment that is suitable for controllable tissue regeneration. To address such a challenge, a new and evolving discipline termed "functional tissue engineering" has been developed [9]. The objective of functional tissue engineering research is to elucidate the role of both biological and biophysical factors in tissue regeneration within scaffolds and to develop guidelines for applying these research results to clinical practices. Functional tissue replacements may require additional exogenous interventions in order to achieve long-term success of the implant system. Figure 8.1 shows functional tissue engineering approaches for clinical applications. The biomedical imaging of the defect will be first taken to determine the desired geometry and material of the implant (e.g., particles, gels, fibrous matrix, ceramic scaffold, etc.) depending on the site/feature of fracture and the type of tissue. Then, the surface modification and incorporation of growth factors/proteins will be conducted to improve the osteoinductive and/or osteoconductive properties of the implant system. On the other hand, osteogenic cells will be taken from either the patient or other donors and then will be sorted, expanded, and seeded in the scaffold to augment its function. Based on this situation, cells can even be genetically modified to serve the purpose. Three regimens

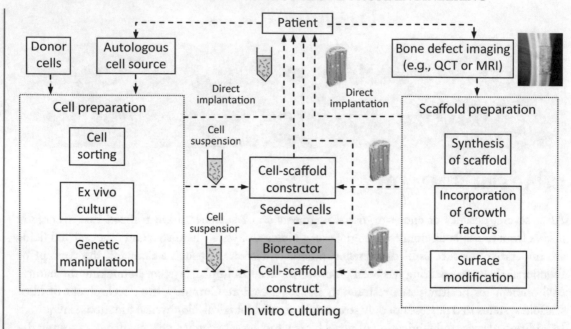

Figure 8.1: Tissue engineering approaches for bone repair using functional scaffold systems.

of implantation are conceivable: (1) direct implantation of cells, (2) implantation of cell seeded constructs, and (3) implantation of in vitro tissue engineered constructs.

It is now well accepted that time-varying changes in stress, strains, fluid pressure/flows, and cell deformation may have influences on the biological activities of normal tissues [5,8,25,64,72,94, 108]. Thus, the success of the cell–scaffold systems in bone tissue engineering is largely dependent on both the macroscopic and microscopic responses of the scaffold system under loads within adequate physiological ranges. In addition, functional tissue engineering of bone cannot be achieved without an adequate combination of mechanical environment and biological/biochemical factors. Understanding of mechanobiology becomes necessary for tissue engineering research, while tissue engineering research, in turn, provides a platform for studying mechanobiology. Previous studies have shown a great progress in research of bone tissue engineering by considering the effect of mechanical forces in the design of the scaffold system. For example, previous studies have shown that mechanical loading can regulate both the tissue properties and microstructural organization [10,15,63].

In this chapter, some current issues pertaining to biomechanics in functional tissue engineering research are presented and discussed. The first important issue in this regard is to understand the mechanobiology of bone, in which biomechanics plays a significant role in the biological activities of bone during the regeneration process. In addition, it is important to develop an ideal scaffold that is biocompatible, bioresorbable, and has structural integrity to provide a framework for cells to grow under appropriate mechanical stimulus. Such a functional scaffold should be able to gradually dissolve away as the newly formed bone is generated while sustaining the desired structural and osteoconductive capability throughout the process. Finally, bioreactors have been utilized to ensure the quality of tissue regeneration *in vitro*. The key functions of bioreactors are two-fold: (1) to provide a controllable mechanical and biological environment during cell/tissue cultures; and (2) to standardize, automate, and scale the manufacturing of tissue engineering products for cost-effective and reproducible clinical applications. Additionally, bioreactors can also serve as a well-defined test platform for controlled research on mechanical stimuli and cellular/molecular functions in three-dimensional environments.

8.2   IDEAL SCAFFOLD SYSTEM FOR BONE TISSUE ENGINEERING

An ideal scaffold system for bone tissue engineering has to be osteoconductive and osteoinductive, be able to sustain the mechanical/structural stability of defect bone, and to dissolve gradually over time until completely transferring the external load to the healed bone tissues. However, none of the current scaffold systems actually meet the requirements. In general, the orthopaedic/dental implant systems can be categorized in two groups: tissue engineered constructs and orthopaedic/dental implants (or grafts). The former requires *in vitro* preconditioning or culturing of cell-scaffold constructs prior to implantation [40, 57, 105, 106], whereas the latter is directly implanted into patients [12, 13, 67, 84]. As shown in Table 8.1, the implant systems can also be grouped in terms of types: allografts based, growth factor based, cell based, ceramics based, and polymer based. Selection of these implant systems is dependent on the nature (e.g., critical size or not) and site (e.g., load bearing or not) of the defect bone.

8.2.1   FUNCTIONAL ARCHITECTURE OF SCAFFOLD

Biomimetics of bone tissue engineering scaffold is essential for functional tissue engineering systems. The architecture of the scaffolds (e.g., pore size, porosity, interconnectivity and permeability) is critical for ensuring mechanical properties of the scaffold systems that are closely-matched to those of native tissues in order to sustain the suitable load transfer necessary to regulate, adapt, and remodel bone during the normal healing process Additionally, the architecture of the scaffold is required to allow for sustained cell proliferation and differentiation within the scaffolds with favorable transport/diffusion of ion, nutrients, and wastes. Obviously, a functional design of scaffold architecture should benefit the later function and restoration of the regenerated tissue.

Currently, many attempts have been made to fabricate scaffold systems that mimic the trabecular bone structures using different biomaterials that are either biodegradable or non biodegradable [6, 27, 58, 83, 101]. One recent progress in this area is the development of the functional bioceramic scaffolds that have a hierarchical architecture at all macro, micro, and nanoscopic levels. As shown in Figure 8.2, scaffolds can be fabricated to have a similar structure of bone tissues (Fig-

Table 8.1: Classification of current implant systems used in bone tissue engineering.

Type	Description	Biodegradability	Examples
Graft based	Natural bone grafts, used alone or in combination with other materials	No	Allogro, OrthoBlast, Opteform, Grafton
Factor based	Natural and recombinant growth factors, used alone or in combination with other materials	Yes	TGF-beta, PDGF, FGF, BMP
Cell based	Cells used to generate new tissue alone or seeded onto a support matrix	Yes	Mesenchymal stem cells
Ceramic based	HA, calcium phosphate, calcium sulfate, and bioglass, used alone or in combination	Yes & No	Osteograf, Norian SRS, ProOsteon, OSTEOSET
Polymer based	Polymers (including collagen), used alone or in combination with other materials	Yes & No	Cortoss, Collagraft OPLA, IMMIX

ure 8.2(a) and (b)) by controlling the local porosity of the scaffold. Moreover, the trabeculae of the scaffold could be made hollow (i.e., microchannels) for ease of transportation of ions and growth factors within the scaffold (Figure 8.2(c)). For further permeation and drug/factor release into the system, the base material of the trabeculae may be made porous at nanoscopic levels, which are permeable for small molecules transportation (Figure 8.2(d)). Additionally, the scaffolds can be

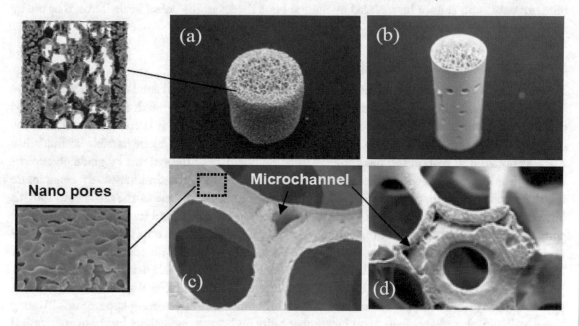

Figure 8.2: Examples of functional tissue engineering scaffold: (a) a scaffold having a structure with low porosity at the shell and high porosity inside to mimic the natural bone structure, (b) a scaffold with a trabecular structure enclosed in a dense shell that has holes for permeation, (c) and (d) the hollow trabeculae having a micro channel inside and having a porous structure at nanoscale levels (all photos are provided as the courtesy by Drs. Oh and Ong at the University of Texas San Antonio).

made as a composite of hydroxyapatite and tri-calcium phosphate or other biomaterials to meet the mechanical requirement (e.g., stiffness and strength) of the scaffold and to control of the biodegradability based on the actual applications. Finally, the scaffold system can serve a drug/factor carrier for modulate the cell activity during the tissue regeneration process.

8.2.2   NON BIODEGRADABLE SCAFFOLDS

Non biodegradable materials, which include metals (e.g., titanium and Co-Cr alloys), ceramics (e.g., Alumina, Zirconia, and HA), and polymers (e.g., Polyethylene) are usually used in total joint replacements [38], orthopaedic/dental implant systems [48,49], and correction of skeletal deformities [99]. From biomechanics perspectives, they have to sustain the similar stiffness and strength as the normal tissues do in order to ensure the mechanical/structural stability of the skeletal tissues to be replaced. In addition, current advances in nanoscience have enabled the treatments for biologically preferred biomaterial surfaces. For instance, bioactive ceramics have been used as coating materials on metallic implants, such as titanium implants coated with calcium phosphate that allow for cell seeding for better bone ingrowth into the surface of the implants [54]. Nonetheless, these materials are not biodegradable and cannot be replaced by natural bone tissues as the defect heals. Thus, they are by no means ideal implant systems.

8.2.3   BIODEGRADABLE SCAFFOLDS

To construct ideal implant systems for bone tissue engineering, biodegradable scaffolds are the indisputable candidates because they allow for induction of bone ingrowth into the systems and can dissolve away completely over time. For load bearing bone repair, it is critical to maintain the consistent mechanical/structural integrity at the healing sites. Use of biodegradable scaffolds has made it possible to sustain the mechanical/structural integrity of the systems by gradually replacing the implant with newly formed bone tissues, which in turn can simultaneously compensate for the mechanical/structural loss induced by implant degradation. One of the challenges for this kind of scaffold is how to maintain a balance between bone ingrowth and implant degradation in a controlled way so that a consistent structural/mechanical integrity of the implant systems can be sustained throughout the healing process of bone defects [103]. Currently, biodegradable bioceramics (e.g., calcium phosphates) [19,23,26,32,81], polymers (e.g., Poly-lactide acid, poly-glycolide acid, polydioxanone) [3,30,55,79], gels (e.g., collagen, hyaluronic acid) [24,89,92,109], and their composites [52,65] have been used to make scaffolds for bone tissue engineering applications. Among them, calcium phosphates have good biocompatibility and osteoconductivity but poor mechanical integrity, which has been studied extensively for improvement [20,70]. Polymer scaffolds with seeded cells and growth factors have been used for repair of non load-bearing bone defects [37]. However, recent studies have proposed different techniques and strategies (e.g., composites) to augment the mechanical integrity of polymer scaffolds that can be used in load bearing bone repair [1,29,51].

8.2.4   OSTEOINDUCTIVITY/OSTEOCONDUCTIVITY

Osteoinductivity and/or osteoconductivity are also critical for ideal scaffold systems. Osteoinductive scaffolds contain growth factors (e.g., BMPs) or other osteoinductive fillers (e.g., autograft particles) capable of independently attracting precursor cells and inducing new bone formation [50]. On the other hand, osteoconductive scaffolds are those that are permissive for bone formation without incorporating growth factors to attract osteoblast precursors and initiate osteoblastic differentiation and bone formation (e.g., collagen, hydroxyapatite/calcium phosphate, organic polymers, and coral, etc.) [61]. There are multiple factors that affect the osteoinductivity and osteoconductivity of scaffolds. For example, biologically preferred surfaces of scaffolds can induce bone cell activities and enhance the ability of bone formation (or osteoinductivity) indexed using alkaline phosphatase concentrations. In addition to the scaffold materials, the architecture is another important factor that affects the osteoinductivity and osteoconductivity of scaffolds. The pore size and permeability of scaffolds may significantly influence the cell migration, nutrient and waste transportation, and biomechanical environment around cells within the scaffolds [11,36].

8.2.5   MECHANICAL ENVIRONMENT

Maintaining an adequate mechanical environment is another challenge in developing functional scaffold systems [34,35,39,46,66,77]. Since both fluid flow [39] and local strain [60] are mechanical stimuli to bone cells, the architecture and porosity may significantly affect the fluid flow behavior and local strains within the scaffold, thereby altering the cell behavior accordingly. In addition, it is reported that microdamage is another biophysical stimulus to cells in bone tissues [62]. Thus, the effect of microdamage formation and progression on seeded cells in the scaffold system also should be investigated. Recently, computer simulations have been used to help determine the mechanical environment of cells when seeded into a scaffold and to ensure the proper design of the geometry and stiffness of the scaffold [44,77]. However, it is by no means an easy task because the functional scaffold is a dynamic system, whose mechanical/structural, biological, and biophysical environment varies constantly due to the complex biochemical (e.g., biomaterial degradation, etc.) cell activities (e.g., differentiation, proliferation, etc.), and biophysical (e.g., fluid flow, local strain, temperature, etc.) processes. It is anticipated that mathematical modeling of tissue regeneration will be the future trend in design and manufacturing of functional tissue engineering products.

8.3   BIOREACTOR: BIOMIMETIC ENVIRONMENT FOR BONE FORMATION

Successful generation of three-dimensional tissue constructs *in vitro* is dependent on two major aspects: new biological models of cell cultures and adequate physicochemical environments of the scaffold system. To create a suitable physicochemical environment for cell cultures, one challenge is the design of scaffolds that could mimic the natural properties of bone while providing a temporary scaffold for tissue regeneration. Currently, bioresorbable implants have been commonly used to serve the purpose. The challenge lies in matching the degradation and loss in mechanical properties with the ingrowth and formation of bone. To resolve the challenges, a bioreactor becomes necessary to optimize the parameters and eventually to produce synthetic bone grafts that match the properties of natural trabecular bone. Considering the requirements of bone tissue engineering, the key functions of a bioreactor is defined as providing adequate control and standardization of the process of producing tissue engineered products [56]. To do so, the bioreactor has to offer a controllable physicochemical environment for the cell-tissue culture systems to produce high quality engineered tissues [97]. In addition, the process has to be standardized, automated, and possibly scaled for manufacturing tissue grafts in order to ensure economical viability and reproducibility. In fact, bioreactors are *in vitro* culture systems that mimic the *in vivo* environment that implanted scaffolds may experience. Bioreactor based bone tissue engineering has become the future direction of tissue engineering

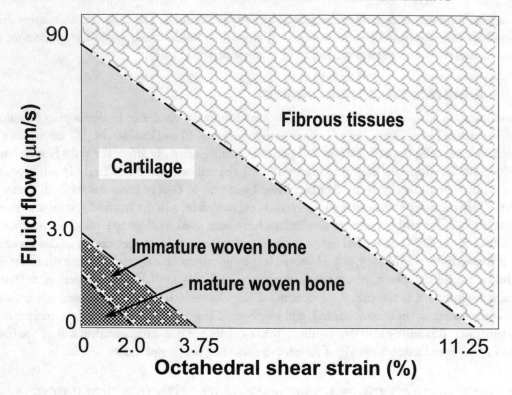

Figure 8.3: Tissue type as a function of fluid flow and applied strain levels (modified based on the data reported in [45]).

research [106]. Many challenges are encountered in the current development of bioreactors. For instance, uniform cell seeding, adequate mass transportation for delivery of nutrients and oxygen and removal of metabolic waste products, proper mechanical environment mimicking cell physiology *in vivo* are all important topics that are still under investigation.

8.3.1   CELL SEEDING

Uniform seeding of cells in scaffolds is an initial but critical step in the generation of functional tissues. It is a major challenge to distribute cells evenly and throughout the scaffold in an effective and reproductive manner. There are several techniques currently used. For example, so called 'static seeding' technique is simply pipetting a cell concentrated suspension into a porous scaffold. This technique obviously lacks control and reproducibility. Stirring can improve the quality and reproducibility of the seeding process when scaffolds are highly porous or have large face area/volume ratios [96]. Perfusion seeding technique is a more efficient and effective approach of evenly distributing cells in scaffolds [100]. However, it is obviously questionable regarding the productivity of the technique for manufacturing the scaffolds that have low porosity and complex pore configurations. Introducing agitation or convective fluid flow into the system can accelerate the uniform seeding of cells, but it may cause adverse outcomes to the stress sensitive cells due to the mechanobiological effects (e.g., mechanical stimuli to cells). Due to the versatility of the scaffold systems responding to different applications, it is still a challenging issue to uniformly distribute cells throughout the different scaffold systems without affecting the cells in harmful ways.

8.3.2 MASS TRANSPORT

It is well known that the cell viability within the interior of the scaffold is the major challenge for the success of tissue engineered bone products. To ensure cell viability, it is required to provide sufficient nutrients and oxygen to the cells, while removing all metabolic waste products. Thus, mass transport both locally and as a whole becomes critical for uniform and healthy bone tissue regeneration in the scaffolds. Compared with convective systems, perfusion bioreactors have been shown to provide better tissue growth and matrix mineralization by bone cells [85]. To balance the mass transport for nutrients and metabolic wastes and the fluid flow in the scaffold system, it has been proposed to use computational fluid dynamics modeling in conjunction with flow visualization techniques to ensure the adequate design of scaffold systems [102]. Ideally, optimization of operating conditions for controllable mass transport in the scaffold system is required and needs to be systemically researched.

8.3.3 PHYSICAL CONDITIONING

Physical environment in the functional scaffold system is also critical for success of tissue regeneration *in vitro*. Compared to temperature, diffusive gradient, and other physical conditions, mechanical conditions (e.g., pressure, stress, strain, microdamage formation, etc) are the most important physical stimuli of tissue regeneration. Recently, more and more studies are performed to elucidate the effect of mechanical stimuli on the tissue regeneration *in vivo* [39, 43, 47, 69, 71, 72, 93] and/or in functional scaffolds [57, 59, 76, 77]. First, the architecture of porous scaffolds (including organization and orientation) directly affects the local pressure, stress, and strain applied to cells seeded into the scaffold system. Next, the fluid flow with the scaffold is also affected by the architecture of the system. Thus, strategic consideration of *mechanical stimulation* in the bioreactor system should be taken during the *in vitro* fabrication process or early post-surgical periods. By controlling the mechanical stimuli (e.g., magnitude and direction of strains or stresses) in the bioreactor, the cells may be instructed to synthesize an appropriately aligned matrix and eventually desired bone tissues. However, more information is required to determine the adequate magnitude and frequency of mechanical stimuli in functional tissue engineering for different bone tissues and anatomic sites.

8.3.4 BIOLOGICAL/BIOCHEMICAL ENVIRONMENT

Functional tissue engineering bioreactors should provide adequate biological environment for cells or cell systems in the scaffold to generate new bone tissues both efficiently and effectively [33,53,87,88, 90]. The most important biological/biochemical conditions include scaffold surface chemistry [7,22] and pH, gas mix, oxygen concentration, glucose concentration, growth factors, etc., in the media. In fact, cell-scaffold surface attachment is a key for the later cell shape, cell proliferation, signal transduction, cell differentiation, cell function, and ultimately the tissue integrity [74, 75, 80]. Like in the healthy human body, balanced constituents in the media have to be provided to the cells. Although numerous studies have been reported on individual or limited biological/biochemical conditions, no systematic studies have been performed to date to address this important issue.

8.4   MECHANOBIOLOGY IN BONE TISSUE ENGINEERING

It has been well accepted in the past decade that mechanical factors should be an integral part of design and manufacturing of tissue engineered bone products [9]. Bioengineers have realized that mechanical stimulation is as equally important as appropriate growth factors to the success of engineered bone tissues. For meaningful design of functional scaffold systems, bioengineers have to know the mechanobiological pathways of such effects. Mechanobiology includes two important aspects: mechanosensation and mechanotransduction. However, the underlying mechanism of such mechanobiological behaviors of bone is still not fully understood.

8.4.1   MECHANOBIOLOGY

From the observation that the bone in the dominating limb of athletes becomes thicker compared with the non-playing one, it has been well accepted that mechanical loading improves bone strength by inducing bone formation in regions where high strain is applied. Currently, two major types of mechanical stimuli to bone/cells (e.g., fluid flow induced shear strain vs. matrix deformation induced local strain) have been identified [14, 60, 82, 86, 98, 107]. However, Figure 8.2 shows that cells will respond differently based on the combination of fluid flow rates and local strain levels, thereby leading to the formation of different tissues [45]. It suggests that the mechanical stimuli have to be in a proper range for cell proliferation and differentiation and eventually bone formation. In the past, extensive studies have been performed to investigate bone responses to different mechanical stimuli; it is still far away from fully understanding the underlying mechanism of the synergy of biological and mechanical processes in the tissue.

8.4.2   MECHANOSENSATION

While it is poorly understood how bone senses the surrounding mechanical environment, osteocytes are considered by most as the mechano-sensors because they are directly attached to the bone matrix and may be loaded by strains induced either by the fluid flow or the local deformation of the extracellular matrix [4, 14, 16, 28, 41, 98]. Since the mechanical interaction between the cell and the extracellular matrix is most likely viscoelastic, the mechanical stimulus to the osteocyte would be very sensitive to the loading rate. This could explain the experimental observation that increasing loading frequency (or rate) improves the responsiveness of bone formation [17]. It is also noteworthy that collagen plays a pivotal role in the delivery of mechanical signals to cells via adhesion between the cells and the substrates [21]. In addition, cellular biology indicates that the production of sclerotic (osteocyte-specific protein) may be suppressed by mechanical loading [73], thus allowing for Wnt signaling-dependent bone formation to occur [78,95]. Moreover, osteocytes may function as mechano-transducers by regulating local osteoclastogenesis via soluble signals [104]. Nonetheless, numerous questions still remain. For example, besides osteocytes are there any other mechano-sensors in bone? One possible candidate is nerve cells since sensory nerve fibers have been identified to be inside bone tissues, especially in Haversian canals [68].

8.4.3   MECHANOTRANSDUCTION

Another important issue for mechanobiology of bone is mechanotransduction, i.e., the communication between the same and different bone cells (e.g., osteocytes, osteoblasts, and osteoclasts) along the pathways of signal transduction for bone formation and resorption [18,31,42]. Using a novel explant system, it is reported that mechanical stimulation enhances the osteocyte viability and osteocytes play a critical role in the transmission of mechanical loading induced signals to modulate osteoblast function [91]. In fact, osteocytes communicate with each other through podocytic extensions (with canaliculi), and with other cells, such as bone–lining cells, osteoblasts, periosteum, and possibly cells in the bone marrow cavity [2]. An activated osteocyte by mechanical stimulus could activate a large amount of surrounding osteocytes through gap junction or paracrine factors released to the canalicular fluid.

Upon fully understanding the mechanobiology of bone, computational modeling becomes necessary to optimize the scaffold design (architecture, material, shape and size, etc.), taking into consideration scaffold degradation, fluid flow and shear stress changes, local deformation induced strains, cellular responses, and tissue regeneration process.

8.5   SUMMARY

The challenge to developing usable tissue engineered products for bone repairs and replacement is how to make them suitable for the *in vivo* environment for the best tissue regeneration outcomes. The biomechanics related to functional bone tissue engineering is how to design and manufacture the tissue engineered constructs to ensure the mechanical stress/strain experienced by the implants are at optimal ranges and orientations so that the best bone repair/replacement can be realized at a variety of anatomic locations.

Clearly, the tissue engineering community needs to establish functional criteria that will help bioengineers to design and manufacture tissue engineered bone products for bone repairs and replacements. By establishing these criteria, new and more innovative tissue engineered repairs and replacements can be provided for clinical applications. From the production perspective, scale-up, packaging, storage, and handling procedures are also critical to bringing the final products to doctors and patients. Briefly, the implants must be capable of retaining their mechanical, structural, and biological integrity during large-scale production, packaging, and storage. In addition, the products have to be easy to handle in the operating room, thus encouraging surgeons to use the devices in patients.

The biomechanical, biophysical, biochemical, and biological environments for tissue engineered systems can be optimized via introduction of novel bioreactors during the fabrication or delivery of the constructs, thus offering the guaranteed quality of tissue engineered products. Ideally, such controls must be temporally and spatially delivered in an appropriate fashion to stimulate early cell proliferation, subsequent matrix synthesis, rapid mineralization, quick adaptation to the *in vivo* environment, and maintenance of proper mechanical/structural integrity for resisting the large forces applied during daily activity. With rapidly evolving new technologies being developed, the future of tissue engineering for bone repair/replacement is quite promising.

REFERENCES

[1] Ahmed, I., et al., Retention of mechanical properties and cytocompatibility of a phosphate-based glass fiber/polylactic acid composite. Journal of biomedical materials research. Part B, Applied biomaterials, 2009. **89**(1): p. 18–27. DOI: 10.1002/jbm.b.31182 8.2.3

[2] Allori, A.C., et al., Biological basis of bone formation, remodeling, and repair-part III: biomechanical forces. Tissue engineering. Part B, Reviews, 2008. **14**(3): p. 285–93. DOI: 10.1089/ten.teb.2008.0084 8.4.3

[3] Alsberg, E., et al., Regulating bone formation via controlled scaffold degradation. Journal of dental research, 2003. **82**(11): p. 903–8. DOI: 10.1177/154405910308201111 8.2.3

[4] Apostolopoulos, C.A. and D.D. Deligianni, Prediction of local cellular deformation in bone–influence of microstructure dimensions. Journal of musculoskeletal & neuronal interactions, 2009. **9**(2): p. 99–108. 8.4.2

[5] Bacabac, R.G., et al., Bone cell responses to high-frequency vibration stress: does the nucleus oscillate within the cytoplasm? FASEB J., 2006. **20**(7): p. 858–864. DOI: 10.1096/fj.05-4966.com 8.1

[6] Boschetti, F., et al., Design, fabrication, and characterization of a composite scaffold for bone tissue engineering. The International journal of artificial organs, 2008. **31**(8): p. 697–707. 8.2.1

[7] Brizzola, S., et al., Morphologic features of biocompatibility and neoangiogenesis onto a biodegradable tracheal prosthesis in an animal model. Interactive cardiovascular and thoracic surgery, 2009. **8**(6): p. 610–4. DOI: 10.1510/icvts.2008.197012 8.3.4

[8] BURGER, E.H. and J. KLEIN-NULEND, Mechanotransduction in bone—role of the lacuno-canalicular network. FASEB J., 1999. **13**(9001): p. 101–112. 8.1

[9] Butler, D.L., S.A. Goldstein, and F. Guilak, Functional Tissue Engineering: The Role of Biomechanics. Journal of biomechanical engineering, 2000. **122**(6): p. 570–575. DOI: 10.1115/1.1318906 8.1, 8.4

[10] Carter, D.R., Mechanical loading histories and cortical bone remodeling. Calcified tissue international, 1984. **36**: p. S19–24. DOI: 10.1007/BF02406129 8.1

[11] Chang, B.-S., et al., Osteoconduction at porous hydroxyapatite with various pore configurations. Biomaterials, 2000. **21**(12): p. 1291–1298. DOI: 10.1016/S0142-9612(00)00030-2 8.2.4

[12] Chang, S.H., et al., Fabrication of pre-determined shape of bone segment with collagen-hydroxyapatite scaffold and autogenous platelet-rich plasma. Journal of materials science. Materials in medicine, 2009. **20**(1): p. 23–31. DOI: 10.1007/s10856-008-3507-1 8.2

[13] Chesnutt, B.M., et al., Design and characterization of a novel chitosan/nanocrystalline calcium phosphate composite scaffold for bone regeneration. Journal of biomedical materials research. Part A, 2009. **88**(2): p. 491–502. DOI: 10.1002/jbm.a.31878 8.2

[14] Cowin, S.C., Mechanosensation and fluid transport in living bone. Journal of musculoskeletal & neuronal interactions, 2002. **2**(3): p. 256–60. 8.4.1, 8.4.2

[15] Datta, N., et al., In vitro generated extracellular matrix and fluid shear stress synergistically enhance 3D osteoblastic differentiation. Proceedings of the National Academy of Sciences of the United States of America, 2006. **103**(8): p. 2488–93. DOI: 10.1073/pnas.0505661103 8.1

[16] Deligianni, D.D. and C.A. Apostolopoulos, Multilevel finite element modeling for the prediction of local cellular deformation in bone. Biomechanics and modeling in mechanobiology, 2008. **7**(2): p. 151–9. DOI: 10.1007/s10237-007-0082-1 8.4.2

[17] Donahue, T.L., et al., Mechanosensitivity of bone cells to oscillating fluid flow induced shear stress may be modulated by chemotransport. Journal of biomechanics, 2003. **36**(9): p. 1363–71. DOI: 10.1016/S0021-9290(03)00118-0 8.4.2

[18] Doyle, A.M., R.M. Nerem, and T. Ahsan, Human mesenchymal stem cells form multicellular structures in response to applied cyclic strain. Annals of biomedical engineering, 2009. **37**(4): p. 783–93. DOI: 10.1007/s10439-009-9644-y 8.4.3

[19] Dyson, J.A., et al., Development of custom-built bone scaffolds using mesenchymal stem cells and apatite-wollastonite glass-ceramics. Tissue engineering, 2007. **13**(12): p. 2891–901. DOI: 10.1089/ten.2007.0124 8.2.3

[20] El-Ghannam, A., Bone reconstruction: from bioceramics to tissue engineering. Expert review of medical devices, 2005. **2**(1): p. 87–101. DOI: 10.1586/17434440.2.1.87 8.2.3

[21] El Haj, A.J., et al., Controlling cell biomechanics in orthopaedic tissue engineering and repair. Pathologie-biologie, 2005. **53**(10): p. 581–9. DOI: 10.1016/j.patbio.2004.12.002 8.4.2

[22] Endres, M., et al., Osteogenic induction of human bone marrow-derived mesenchymal progenitor cells in novel synthetic polymer-hydrogel matrices. Tissue engineering, 2003. **9**(4): p. 689–702. DOI: 10.1089/107632703768247386 8.3.4

[23] Flatley, T.J., K.L. Lynch, and M. Benson, Tissue response to implants of calcium phosphate ceramic in the rabbit spine. Clinical orthopaedics and related research, 1983(179): p. 246–52. 8.2.3

[24] Gao, C., et al., Characteristics of calcium sulfate/gelatin composite biomaterials for bone repair. Journal of biomaterials science. Polymer edition, 2007. **18**(7): p. 799–824. DOI: 10.1163/156856207781367710 8.2.3

[25] García-López, S., et al., Mechanical deformation inhibits IL-10 and stimulates IL-12 production by mouse calvarial osteoblasts in vitro. Archives of oral biology, 2005. **50**(4): p. 449–452. DOI: 10.1016/j.archoralbio.2004.09.001 8.1

[26] Gatti, A.M., et al., Bone augmentation with bioactive glass in three cases of dental implant placement. Journal of biomaterials applications, 2006. **20**(4): p. 325–39. DOI: 10.1177/0885328206054534 8.2.3

[27] Gupta, D., et al., Nanostructured biocomposite substrates by electrospinning and electrospraying for the mineralization of osteoblasts. Biomaterials, 2009. **30**(11): p. 2085–94. DOI: 10.1016/j.biomaterials.2008.12.079 8.2.1

[28] Han, Y., et al., Mechanotransduction and strain amplification in osteocyte cell processes. Proceedings of the National Academy of Sciences of the United States of America, 2004. **101**(47): p. 16689–94. DOI: 10.1073/pnas.0407429101 8.4.2

[29] Hasirci, V., et al., Versatility of biodegradable biopolymers: degradability and an in vivo application. Journal of biotechnology, 2001. **86**(2): p. 135–50. DOI: 10.1016/S0168-1656(00)00409-0 8.2.3

[30] Hedberg, E.L., et al., In vivo degradation of porous poly(propylene fumarate)/poly(DL-lactic-co-glycolic acid) composite scaffolds. Biomaterials, 2005. **26**(22): p. 4616–23. DOI: 10.1016/j.biomaterials.2004.11.039 8.2.3

[31] Hoffler, C.E., et al., Novel explant model to study mechanotransduction and cell-cell communication. Journal of orthopaedic research : official publication of the Orthopaedic Research Society, 2006. **24**(8): p. 1687–98. DOI: 10.1002/jor.20207 8.4.3

[32] Hoogendoorn, H.A., et al., Long-term study of large ceramic implants (porous hydroxyapatite) in dog femora. Clinical orthopaedics and related research, 1984(187): p. 281–8. 8.2.3

[33] Howard, D., et al., Tissue engineering: strategies, stem cells and scaffolds. Journal of anatomy, 2008. **213**(1): p. 66–72. DOI: 10.1111/j.1469-7580.2008.00878.x 8.3.4

[34] Isaksson, H., et al., Bone regeneration during distraction osteogenesis: mechano-regulation by shear strain and fluid velocity. Journal of biomechanics, 2007. **40**(9): p. 2002–11. DOI: 10.1016/j.jbiomech.2006.09.028 8.2.5

[35] Isaksson, H., et al., Comparison of biophysical stimuli for mechano-regulation of tissue differentiation during fracture healing. Journal of biomechanics, 2006. **39**(8): p. 1507–16. DOI: 10.1016/j.jbiomech.2005.01.037 8.2.5

[36] Jones, A.C., et al., The correlation of pore morphology, interconnectivity and physical properties of 3D ceramic scaffolds with bone ingrowth. Biomaterials, 2009. **30**(7): p. 1440–51. DOI: 10.1016/j.biomaterials.2008.10.056 8.2.4

[37] Kanczler, J.M., et al., The effect of mesenchymal populations and vascular endothelial growth factor delivered from biodegradable polymer scaffolds on bone formation. Biomaterials, 2008. **29**(12): p. 1892–900. DOI: 10.1016/j.biomaterials.2007.12.031 8.2.3

[38] Katti, K.S., Biomaterials in total joint replacement. Colloids and surfaces. Biointerfaces, 2004. **39**(3): p. 133–42. DOI: 10.1016/j.colsurfb.2003.12.002 8.2.2

[39] Kavlock, K.D. and A.S. Goldstein, Effect of pulsatile flow on the osteogenic differentiation of bone marrow stromal cells in porous scaffolds. Biomedical sciences instrumentation, 2008. **44**: p. 471–6. 8.2.5, 8.3.3

[40] Kempen, D.H., et al., Effect of autologous bone marrow stromal cell seeding and bone morphogenetic protein-2 delivery on ectopic bone formation in a microsphere/poly(propylene fumarate) composite. Tissue engineering. Part A, 2009. **15**(3): p. 587–94. DOI: 10.1089/ten.tea.2007.0376 8.2

[41] Klein-Nulend, J., R.G. Bacabac, and M.G. Mullender, Mechanobiology of bone tissue. Pathologie-biologie, 2005. **53**(10): p. 576–80. DOI: 10.1016/j.patbio.2004.12.005 8.4.2

[42] Ko, K.S. and C.A. McCulloch, Intercellular mechanotransduction: cellular circuits that coordinate tissue responses to mechanical loading. Biochemical and biophysical research communications, 2001. **285**(5): p. 1077–83. DOI: 10.1006/bbrc.2001.5177 8.4.3

[43] Kuruvilla, S.J., et al., Site specific bone adaptation response to mechanical loading. Journal of musculoskeletal & neuronal interactions, 2008. **8**(1): p. 71–8. 8.3.3

[44] Lacroix, D., J.A. Planell, and P.J. Prendergast, Computer-aided design and finite-element modelling of biomaterial scaffolds for bone tissue engineering. Series A, Mathematical, physical, and engineering sciences, 2009. **367**(1895): p. 1993–2009. 8.2.5

[45] Lacroix, D., et al., Biomechanical model to simulate tissue differentiation and bone regeneration: application to fracture healing. Medical & biological engineering & computing, 2002. **40**(1): p. 14–21. DOI: 10.1007/BF02347690 8.3, 8.4.1

[46] Lamerigts, N.M., et al., Incorporation of morsellized bone graft under controlled loading conditions. A new animal model in the goat. Biomaterials, 2000. **21**(7): p. 741–7. DOI: 10.1016/S0142-9612(99)00247-1 8.2.5

[47] Lee, T.C., A. Staines, and D. Taylor, Bone adaptation to load: microdamage as a stimulus for bone remodelling. Journal of anatomy, 2002. **201**(6): p. 437–46. DOI: 10.1046/j.1469-7580.2002.00123.x 8.3.3

[48] Lemons, J.E., Dental implant biomaterials. Journal of the American Dental Association (1939), 1990. **121**(6): p. 716–9. 8.2.2

[49] Lemons, J.E., Biomaterials, biomechanics, tissue healing, and immediate-function dental implants. The Journal of oral implantology, 2004. **30**(5): p. 318–24. DOI: 10.1563/0712.1 8.2.2

[50] Lewandrowski, K.-U., et al., Quantitative Measures of Osteoinductivity of a Porous Poly(propylene fumarate) Bone Graft Extender. Tissue engineering, 2003. **9**(1): p. 85–93. 8.2.4

[51] Lewandrowski, K.U., et al., Composite resorbable polymer/hydroxylapatite composite screws for fixation of osteochondral osteotomies. Bio-medical materials and engineering, 2002. **12**(4): p. 423–38. 8.2.3

[52] Liao, S.S., et al., Hierarchically biomimetic bone scaffold materials: nano-HA/collagen/PLA composite. Journal of biomedical materials research. Part B, Applied biomaterials, 2004. **69**(2): p. 158–65. 8.2.3

[53] Lickorish, D., L. Guan, and J.E. Davies, A three-phase, fully resorbable, polyester/calcium phosphate scaffold for bone tissue engineering: Evolution of scaffold design. Biomaterials, 2007. **28**(8): p. 1495–502. DOI: 10.1016/j.biomaterials.2006.11.025 8.3.4

[54] Lopez-Heredia, M.A., et al., Rapid prototyped porous titanium coated with calcium phosphate as a scaffold for bone tissue engineering. Biomaterials, 2008. **29**(17): p. 2608–15. DOI: 10.1016/j.biomaterials.2008.02.021 8.2.2

[55] Low, S.W., et al., Use of Osteoplug polycaprolactone implants as novel burr-hole covers. Singapore medical journal, 2009. **50**(8): p. 777–80. 8.2.3

[56] Martin, I., D. Wendt, and M. Heberer, The role of bioreactors in tissue engineering. Trends in Biotechnology, 2004. **22**(2): p. 80–86. DOI: 10.1016/j.tibtech.2003.12.001 8.3

[57] Martins, A.M., et al., Natural stimulus responsive scaffolds/cells for bone tissue engineering: influence of lysozyme upon scaffold degradation and osteogenic differentiation of cultured marrow stromal cells induced by CaP coatings. Tissue engineering. Part A, 2009. **15**(8): p. 1953–63. DOI: 10.1089/ten.tea.2008.0023 8.2, 8.3.3

[58] Mastrogiacomo, M., et al., Tissue engineering of bone: search for a better scaffold. Orthodontics & craniofacial research, 2005. **8**(4): p. 277–84. DOI: 10.1111/j.1601-6343.2005.00350.x 8.2.1

[59] Maul, T.M., et al., A new experimental system for the extended application of cyclic hydrostatic pressure to cell culture. Journal of biomechanical engineering, 2007. **129**(1): p. 110–6. DOI: 10.1115/1.2401190 8.3.3

[60] McGarry, J.G., et al., A comparison of strain and fluid shear stress in stimulating bone cell responses–a computational and experimental study. The FASEB journal : official publication of the Federation of American Societies for Experimental Biology, 2005. **19**(3): p. 482–4. DOI: 10.1096/fj.04-2210fje 8.2.5, 8.4.1

[61] McKee, M.D., Management of Segmental Bony Defects: The Role of Osteoconductive Orthobiologics. The Journal of the American Academy of Orthopaedic Surgeons, 2006. **14**(10): p. S163–167. 8.2.4

[62] McNamara, L.M. and P.J. Prendergast, Bone remodelling algorithms incorporating both strain and microdamage stimuli. Journal of biomechanics, 2007. **40**(6): p. 1381–91. DOI: 10.1016/j.jbiomech.2006.05.007 8.2.5

[63] Meinel, L., et al., Bone tissue engineering using human mesenchymal stem cells: effects of scaffold material and medium flow. Annals of biomedical engineering, 2004. **32**(1): p. 112–22. DOI: 10.1023/B:ABME.0000007796.48329.b4 8.1

[64] Nerem, R.M. and A. Sambanis, Tissue Engineering: From Biology to Biological Substitutes. Tissue engineering, 1995. **1**(1): p. 3–13. DOI: 10.1089/ten.1995.1.3 8.1

[65] Ngiam, M., et al., The fabrication of nano-hydroxyapatite on PLGA and PLGA/collagen nanofibrous composite scaffolds and their effects in osteoblastic behavior for bone tissue engineering. Bone, 2009. **45**(1): p. 4–16. DOI: 10.1016/j.bone.2009.03.674 8.2.3

[66] Nishimura, H., et al., Lectins induce resistance to proteases and/or mechanical stimulus in all examined cells–including bone marrow mesenchymal stem cells–on various scaffolds. Experimental cell research, 2004. **295**(1): p. 119–27. DOI: 10.1016/j.yexcr.2003.12.018 8.2.5

[67] Niu, X., et al., Porous nano-HA/collagen/PLLA scaffold containing chitosan microspheres for controlled delivery of synthetic peptide derived from BMP-2. Journal of controlled release : official journal of the Controlled Release Society, 2009. **134**(2): p. 111–7. 8.2

[68] Nixon, A.J. and J.F. Cummings, Substance P immunohistochemical study of the sensory innervation of normal subchondral bone in the equine metacarpophalangeal joint. American journal of veterinary research, 1994. **55**(1): p. 28–33. 8.4.2

[69] Nowlan, N.C., P. Murphy, and P.J. Prendergast, A dynamic pattern of mechanical stimulation promotes ossification in avian embryonic long bones. Journal of biomechanics, 2008. **41**(2): p. 249–58. DOI: 10.1016/j.jbiomech.2007.09.031 8.3.3

[70] Ramay, H.R. and M. Zhang, Biphasic calcium phosphate nanocomposite porous scaffolds for load-bearing bone tissue engineering. Biomaterials, 2004. **25**(21): p. 5171–80. DOI: 10.1016/j.biomaterials.2003.12.023 8.2.3

[71] Rangaswami, H., et al., Type II cGMP-dependent protein kinase mediates osteoblast mechanotransduction. The Journal of biological chemistry, 2009. **284**(22): p. 14796–808. DOI: 10.1074/jbc.M806486200 8.3.3

[72] Robling, A.G., et al., Improved bone structure and strength after long-term mechanical loading is greatest if loading is separated into short bouts. Journal of bone and mineral research : the official journal of the American Society for Bone and Mineral Research, 2002. **17**(8): p. 1545–54. 8.1, 8.3.3

[73] Robling, A.G., et al., Mechanical stimulation of bone in vivo reduces osteocyte expression of Sost/sclerostin. The Journal of biological chemistry, 2008. **283**(9): p. 5866–75. DOI: 10.1074/jbc.M705092200 8.4.2

[74] Ruoslahti, E., Stretching is good for a cell. Science (New York, N.Y.), 1997. **276**(5317): p. 1345–6. 8.3.4

[75] Ruoslahti, E. and M.D. Pierschbacher, New perspectives in cell adhesion: RGD and integrins. Science (New York, N.Y.), 1987. **238**(4826): p. 491–7. DOI: 10.1126/science.2821619 8.3.4

[76] Santos, M.I., et al., Endothelial cell colonization and angiogenic potential of combined nano- and micro-fibrous scaffolds for bone tissue engineering. Biomaterials, 2008. **29**(32): p. 4306–13. DOI: 10.1016/j.biomaterials.2008.07.033 8.3.3

[77] Sanz-Herrera, J.A., J.M. Garcia-Aznar, and M. Doblare, A mathematical model for bone tissue regeneration inside a specific type of scaffold. Biomechanics and modeling in mechanobiology, 2008. **7**(5): p. 355–66. DOI: 10.1007/s10237-007-0089-7 8.2.5, 8.3.3

[78] Sawakami, K., et al., The Wnt co-receptor LRP5 is essential for skeletal mechanotransduction but not for the anabolic bone response to parathyroid hormone treatment. The Journal of biological chemistry, 2006. **281**(33): p. 23698–711. DOI: 10.1074/jbc.M601000200 8.4.2

[79] Schantz, J.T., et al., Repair of calvarial defects with customised tissue-engineered bone grafts II. Evaluation of cellular efficiency and efficacy in vivo. Tissue engineering, 2003. **9**: p. S127–39. DOI: 10.1089/10763270360697030 8.2.3

[80] Schneider, G. and K. Burridge, Formation of focal adhesions by osteoblasts adhering to different substrata. Experimental cell research, 1994. **214**(1): p. 264–9. DOI: 10.1006/excr.1994.1257 8.3.4

[81] Seitz, H., et al., Three-dimensional printing of porous ceramic scaffolds for bone tissue engineering. Journal of biomedical materials research. Part B, Applied biomaterials, 2005. **74**(2): p. 782–8. DOI: 10.1002/jbm.b.30291 8.2.3

[82] Sharp, L.A., Y.W. Lee, and A.S. Goldstein, Effect of low-frequency pulsatile flow on expression of osteoblastic genes by bone marrow stromal cells. Annals of biomedical engineering, 2009. **37**(3): p. 445–53. DOI: 10.1007/s10439-008-9632-7 8.4.1

[83] Shimko, D.A. and E.A. Nauman, Development and characterization of a porous poly(methyl methacrylate) scaffold with controllable modulus and permeability. Journal of biomedical materials research. Part B, Applied biomaterials, 2007. **80**(2): p. 360–9. DOI: 10.1002/jbm.b.30605 8.2.1

[84] Shin, M., H. Yoshimoto, and J.P. Vacanti, In vivo bone tissue engineering using mesenchymal stem cells on a novel electrospun nanofibrous scaffold. Tissue engineering, 2004. **10**(1–2): p. 33-41. DOI: 10.1089/107632704322791673 8.2

[85] Sikavitsas, V.I., et al., Flow perfusion enhances the calcified matrix deposition of marrow stromal cells in biodegradable nonwoven fiber mesh scaffolds. Annals of biomedical engineering, 2005. **33**(1): p. 63–70. DOI: 10.1007/s10439-005-8963-x 8.3.2

[86] Simmons, C.A., et al., Cyclic strain enhances matrix mineralization by adult human mesenchymal stem cells via the extracellular signal-regulated kinase (ERK1/2) signaling pathway. Journal of biomechanics, 2003. **36**(8): p. 1087–96. DOI: 10.1016/S0021-9290(03)00110-6 8.4.1

[87] Soltan, M., D. Smiler, and J.H. Choi, Bone marrow: orchestrated cells, cytokines, and growth factors for bone regeneration. Implant dentistry, 2009. **18**(2): p. 132–41. DOI: 10.1097/ID.0b013e3181990e75 8.3.4

[88] Stiehler, M., et al., Effect of dynamic 3-D culture on proliferation, distribution, and osteogenic differentiation of human mesenchymal stem cells. Journal of biomedical materials research. Part A, 2009. **89**(1): p. 96–107. 8.3.4

[89] Sun, B., et al., Crosslinking heparin to collagen scaffolds for the delivery of human platelet-derived growth factor. Journal of biomedical materials research. Part B, Applied biomaterials, 2009. **91**(1): p. 366–72. DOI: 10.1002/jbm.b.31411 8.2.3

[90] Tabata, Y., Tissue regeneration based on growth factor release. Tissue engineering, 2003. **9**: p. S5–15. DOI: 10.1089/10763270360696941 8.3.4

[91] Takai, E., et al., Osteocyte viability and regulation of osteoblast function in a 3D trabecular bone explant under dynamic hydrostatic pressure. Journal of bone and mineral research : the official journal of the American Society for Bone and Mineral Research, 2004. **19**(9): p. 1403–10. 8.4.3

[92] Trombi, L., et al., Good manufacturing practice-grade fibrin gel is useful as a scaffold for human mesenchymal stromal cells and supports in vitro osteogenic differentiation. Transfusion, 2008. **48**(10): p. 2246–51. DOI: 10.1111/j.1537-2995.2008.01829.x 8.2.3

[93] Tsubota, K., T. Adachi, and Y. Tomita, Functional adaptation of cancellous bone in human proximal femur predicted by trabecular surface remodeling simulation toward uniform stress state. Journal of biomechanics, 2002. **35**(12): p. 1541–51. DOI: 10.1016/S0021-9290(02)00173-2 8.3.3

[94] Turner, C., M. Forwood, and M. Otter, Mechanotransduction in bone: do bone cells act as sensors of fluid flow? FASEB J., 1994. **8**(11): p. 875–878. 8.1

[95] Turner, C.H., et al., Mechanobiology of the skeleton. Science signaling, 2009. **2**(68): p. pt3. DOI: 10.1126/scisignal.268pt3 8.4.2

[96] Vunjak-Novakovic, G., et al., Bioreactor cultivation conditions modulate the composition and mechanical properties of tissue-engineered cartilage. Journal of orthopaedic research : official publication of the Orthopaedic Research Society, 1999. **17**(1): p. 130–8. DOI: 10.1002/jor.1100170119 8.3.1

[97] Wang, S., et al., Vertical alveolar ridge augmentation with beta-tricalcium phosphate and autologous osteoblasts in canine mandible. Biomaterials, 2009. **30**(13): p. 2489–98. DOI: 10.1016/j.biomaterials.2008.12.067 8.3

[98] Weinbaum, S., P. Guo, and L. You, A new view of mechanotransduction and strain amplification in cells with microvilli and cell processes. Biorheology, 2001. **38**(2–3): p. 119-42. 8.4.1, 8.4.2

[99] Weinzweig, J., et al., Osteochondral reconstruction of a non-weight-bearing joint using a high-density porous polyethylene implant. Plastic and reconstructive surgery, 2000. **106**(7): p. 1547–54. DOI: 10.1097/00006534-200012000-00016 8.2.2

[100] Wendt, D., et al., Oscillating perfusion of cell suspensions through three-dimensional scaffolds enhances cell seeding efficiency and uniformity. Biotechnology and bioengineering, 2003. **84**(2): p. 205–14. DOI: 10.1002/bit.10759 8.3.1

[101] Wettergreen, M.A., et al., Computer-aided tissue engineering of a human vertebral body. Annals of biomedical engineering, 2005. **33**(10): p. 1333–43. DOI: 10.1007/s10439-005-6744-1 8.2.1

[102] Williams, K.A., S. Saini, and T.M. Wick, Computational fluid dynamics modeling of steady-state momentum and mass transport in a bioreactor for cartilage tissue engineering. Biotechnology progress, 2002. **18**(5): p. 951–63. DOI: 10.1021/bp020087n 8.3.2

[103] Yeo, A., et al., The degradation profile of novel, bioresorbable PCL-TCP scaffolds: an in vitro and in vivo study. Journal of biomedical materials research. Part A, 2008. **84**(1): p. 208–18 DOI: 10.1002/jbm.a.31454 8.2.3

[104] You, L., et al., Osteocytes as mechanosensors in the inhibition of bone resorption due to mechanical loading. Bone, 2008. **42**(1): p. 172–9. DOI: 10.1016/j.bone.2007.09.047 8.4.2

[105] Yu, H., et al., Improved tissue-engineered bone regeneration by endothelial cell mediated vascularization. Biomaterials, 2009. **30**(4): p. 508–17. DOI: 10.1016/j.biomaterials.2008.09.047 8.2

[106] Yu, X., et al., Bioreactor-based bone tissue engineering: the influence of dynamic flow on osteoblast phenotypic expression and matrix mineralization. Proceedings of the National Academy of Sciences of the United States of America, 2004. **101**(31): p. 11203–8. DOI: 10.1073/pnas.0402532101 8.2, 8.3

[107] Yuge, L., et al., Physical stress by magnetic force accelerates differentiation of human osteoblasts. Biochemical and biophysical research communications, 2003. **311**(1): p. 32–8. DOI: 10.1016/j.bbrc.2003.09.156 8.4.1

[108] Zhang, C., et al., Direct compression as an appropriately mechanical environment in bone tissue reconstruction in vitro. Medical hypotheses, 2006. **67**(6): p. 1414–1418. DOI: 10.1016/j.mehy.2006.05.044 8.1

[109] Zhao, J., et al., Bone regeneration using collagen type I vitrigel with bone morphogenetic protein-2. Journal of bioscience and bioengineering, 2009. **107**(3): p. 318–23 DOI: 10.1016/j.jbiosc.2008.10.007 8.2.3

Authors' Biographies

XIAODU WANG

Dr. Xiaodu Wang is Professor of Mechanical & Biomedical Engineering at the University of Texas at San Antonio (UTSA). He obtained his Doctor degree in Mechanical Engineering and Materials Science at Yokohama National University, Japan in 1990. Prior to joining UTSA in 1999, Dr. Wang was an Assistant Professor (Research) of Orthopaedic Bioengineering in the Orthopaedic Department at the University of Texas Health Science Center at San Antonio. His current research interest is in the areas of biological tissue mechanics, bone remodeling, and nano biomechanics.

JEFFRY S. NYMAN

Dr. Jeffry S. Nyman is a Research Scientist at the Tennessee Valley Healthcare System within the U.S. Department of Veterans Affairs. His primary academic appointment, Research Assistant Professor, is in the Department of Orthopaedics and Rehabilitation at Vanderbilt Medical Center with a secondary appointment in the Department of Biomedical Engineering at Vanderbilt University. Working with his colleagues at the Vanderbilt Center for Bone Biology, Dr. Nyman investigates the role of certain genes (e.g., ATF4, BMP-2, MMP-2 and -9, and NF1) in regulating the inherent quality of bone tissue. This work is a continuation of his graduate studies at the University of California, Davis, and post-doctoral studies at UTSA in bone adaptation and age-related changes in bone toughness.

XUANLIANG DONG

Dr. Xuanliang Dong is currently a Research Assistant Professor in the Department of Mechanical Engineering at UTSA. Dr. Dong got his Ph.D. degree in mechanical engineering from Columbia University in New York City and did his post-doctoral fellowship at Henry Ford Health System in Detroit and University of California at Davis. His current research area is in orthopaedic biomechanics.

HUIJIE LENG

Dr. Huijie Leng is currently an Associate Professor in the Department of Orthopaedics at Peking University Third Hospital, Beijing, China. Dr. Leng received his Ph.D. degree in the mechanical degree from The University of Notre Dame, USA. He did his post-doctoral research at the University of Texas at San Antonio, USA. Currently Dr. Leng is focusing on the research of clinical biomechanics and mechanics of biomaterials.

MICHAEL J. REYES

Michael J. Reyes, M.S.M.E, is a Ph.D. Candidate and Graduate Research Assistant in the Biomedical Engineering Department of UTSA, working on the project that studies age-related changes in bone remodeling and its effect on the mechanical and compositional properties of bone. Before joining UTSA, Mr. Reyes worked as a research engineer in an engineering consulting firm and earned experiences in the design and testing of engineering systems and products.

Printed in the United States
by Baker & Taylor Publisher Services